Dennis Barker is the author of the two previous self-contained volumes of 'The People of the Forces Trilogy', *Soldiering On* and *Ruling the Waves*, of a vivid study of how a threatened aristocratic country estate was saved, *One Man's Estate*, and of other non-fiction books.

He is also the author of three novels and has been a frequent broadcaster. He has been on the staff of the *Guardian* since 1963 as reporter, feature writer, media correspondent and columnist.

Also by Dennis Barker in Sphere Books in 'The People of the Forces Trilogy':

SOLDIERING ON

RULING THE WAVES

DENNIS BARKER

Guarding the Skies

SPHERE BOOKS LIMITED

A SPHERE Book

First published in Great Britain in 1989 by Viking,
a Division of Penguin Books
This edition published in 1990 by Sphere Books Ltd

Printed and bound in Great Britain by
Cox & Wyman Ltd, Reading

ISBN 0 7474 0043 1

Sphere Books Ltd
A Division of
Macdonald & Co (Publishers) Ltd
Orbit House
1 New Fetter Lane
London EC4A 1AR
A member of Maxwell Macmillan Pergamon Publishing Corporation

Contents

PART IV: PUBLIC CONTACTS

Introduction

Civilians in uniform, snooty British Army people sometimes call them.

Bureaucrats of the skies, Royal Navy flyers sometimes pronounce – as a put-down that works, at least to their own satisfaction, if not to that of the Royal Air Force men and women themselves.

They tend to see themselves largely as thinking hi-tech professionals in the flexible craft of defending and attacking in the air: fully as expert as technical professionals would be in civilian life, but having that extra discipline necessary in a fighting service, where lives can be directly at stake.

It would be easy, but probably mistaken, to believe that everyone in the RAF is a pilot manqué, and satisfied or frustrated in proportion to his success in carrying out that ambition. It is certainly true that the RAF contains this anomaly which can have very human consequences: while its whole point is the flying of aircraft, this job is in practice done by only a tiny majority of its members. The rest must see themselves, and be seen, as back-up. This can be irksome when very young pilots – known affectionately in the trade as 'baby budgies' – are tempted to throw their weight about. In practice they don't do it for long. If they try to, they find that the men on the ground have a few little tricks up their professional sleeves.

'Pilots think they are the lords of creation – and so they are, when they are actually flying,' one administrative officer told me. 'But if they get above themselves with *us*, wanting instant attention, we teach them manners. Hold up their travel documents for a few days, until they get really worried. Forget to send their pay. Any pilot who thinks he can automatically go to the head of every queue gets his wings clipped.'

This officer told me he was perfectly happy with the status quo, and had absolutely no desire to be a pilot. Why, after observing and suffering the panache and speed with which he drove his car, did I come to the conclusion that perhaps his disavowal of any jealousy of pilots was not *entirely* to be taken at face value?

This is a book about the *people* of the Royal Air Force. Machinery is the usual focus of books which, unlike this one, are written by experts for experts. I have looked at the RAF as human beings working machinery, rather than as machinery being worked by human beings. I have encountered one or two officers in the forces (very few in the RAF) who did not see the point of writing about serving men and women as people. How anyone can maintain this position after the Falklands campaign, in which professional British officers and men fought good professional Argentinian officers commanding conscript men, I cannot presume to guess.

Any force depends mightily on the character of its professionalism; and that in turn depends on the character of the people possessing it. Yet the usual literature about the forces, written by insiders for insiders, suggests – if only by implication – that armed strength depends upon machinery, which happens to need those faceless creatures called human beings to operate it. In this way all three services tend to be dehumanized in the public mind – the more so the more sophisticated the technology being used. And of all three services, the RAF has arguably the most sophisticated technology – though fortunately it also has some awareness that human character still counts for a lot in the use of it. A force dehumanized in the public mind is not in the best position to justify itself to that public which, of course, consists of the very taxpayers who are standing the cost of both the machinery and the service people's pay.

There are about 87,000 personnel in the RAF – scarcely more than the number of RAF personnel killed or injured in the Second World War. There are 12,700 male officers and 800 female, 68,000 male airmen and 5,000 female. There are about seven hundred aircraft, ranging from Hawk modern jet trainers via VC10 and TriStar air tankers and Chinook and Wessex helicopters to the twin-seat Tornado, the most speedy and flexible way of defending United Kingdom shores or British aircraft anywhere in the world.

The range is still prodigious, in terms both of types of aircraft and of the areas of the world they cover. For Europeans, the most important front line is Germany, where East directly confronts West. Here I found some of the RAF's 10,690 resident personnel inside their horizontal-take-off Harriers at RAF Gütersloh, and quite prepared to talk about them, because the Harriers have no nuclear capability; or inside their Tornados at RAF Brüggen, displaying considerable conversational caution, since the Tornado *does* have a nuclear capability, though the presence or absence of nuclear weapons is *never* discussed.

There are helicopters in Hong Kong to help the civil power keep order and to keep out illegal Chinese immigrants, though they will presumably not stay long once Hong Kong is handed back to the Chinese in 1997. There are the Tornados and other aircraft I saw in Cyprus, the British intelligence centre hardly more than a stone's throw from Beirut, which is vital to the task of watching the Middle East and parts of the Soviet Union.

These are the continuously based forces. There is practically nowhere in the world that could not be reached, even in these penny-pinching times, by an RAF which is severely depleted in numbers since its Second World War peak. Air-to-air refuelling is now a routine science, which was sharpened up under the pressures of the campaign to liberate the Falkland Islands in 1982. It also enables UK-based fighters guarding United Kingdom skies to stay in the air, hounding the enemy all the time, instead of making themselves vulnerable by landing to refuel.

'Flexibility' is certainly the 'in' word in the RAF today. It is all too understandable that it should be so. RAF officers, given half a chance, will spell out to you the reason. It is this. When reacting to events in a different part of the world, on the land you may be talking about days, on the sea you may be talking about hours, but in the air you are probably talking about minutes. Civilians used to taking holiday air flights readily understand this point. What is not so commonly understood is the effect this has on the psychology of RAF men and women: the game is too fast and too vast for hidebound military diktat and slavish obedience. In this respect, if in no other, the men and women of the RAF are indeed 'civilians in uniform'. They are taught to think, and *expected* to think, not merely to obey.

With fighters costing £20 million or so, and pilots costing about £3 million to train, there is no room for futile Charges of the Light Brigade in which slavish and suicidal obedience to arthritic orders is allowed totally to dominate a questioning common sense.

This is the successor volume to my two previous books about the people of the British Army, *Soldiering On*, and the people of the Royal Navy, *Ruling the Waves*. It completes my trio of books, *The People of the Forces Trilogy*. The task and adventure has taken me to the Arctic Circle with the Royal Marines, resisting feigned Russian invasions; to the steamy heat and rains of the Hong Kong–Chinese border, chasing illegal immigrants with the British Army; to the even steamier heat of the Belize jungles in Central America, training to stave off attacks from neighbouring Guatemala; to the more moderate (meteorologically speaking) climate of Cyprus, just east of Suez, credited with being the cradle of Christianity and the conduit of most present-day terrorist activities in the Middle East, seeing how the RAF acts as the eyes and ears of the West in what will probably be the most volatile area of the world for the foreseeable future; and to many other theatres of action, including the nearest front lines in West Germany and Berlin, each with a distinctive and vital character of its own.

The role and importance of the present-day armed forces are not readily understood by the average civilian. To my knowledge, I am the first civilian layman to be permitted to talk to, and quote freely, the men and women of all ranks in all three armed services for such a trilogy. Kipling undoubtedly *would* have been; but his factual books on the subject came before the RAF was formed in 1918 as the successor to the Royal Flying Corps, established in 1912.

The tone of the times is now quite different. So is the tone of my books, which, I hope, are free of a jingoism that would now ring very hollowly. But I hope I have presented a sympathetic if questioning portrait of the men and women of action who comprise British defence forces in an era which is not likely to see future British conquest of other nations but which could conceivably see a conquest of *Britain* becoming an increasing danger. Nor need the danger come only from direct military action: weak forces, weakly led and with second-class citizens in the ranks, could easily be a factor in diplomatic pressure and blackmail, even if direct military adventurism is permanently ruled out by those who may cast avaricious eyes in the direction of the British Isles. But why should we, why should *any* nation, assume that aggressive military action against it is permanently ruled out? Those who profess that they would feel safer from attack if our defences were less 'provocative' to a potential enemy overlook the fact that fear is not the only reason countries are attacked. They are as often attacked out of greed, out of a desire for an extension of power and influence. And even those who claim that pacifism can defuse fear would be hard put to it to explain how it can deflect greed in any field, let alone international affairs. If being unarmed automatically disarmed greed, no unarmed man or woman would ever be mugged in the street. In fact, as we know to our bitter cost, the more vulnerable and defenceless human beings seem, through age or infirmity, the *more* likely they are to become the victim of an attacker.

International affairs are often individual human encounters writ large. The moral should be apparent to the fair-minded, *especially* if they hate war.

The protective man of action as distinct from communication (no one has denied the talking classes profuse public credit in our media-dominated age) has a continuing role, even in a changed world in which peace is in everyone's rational interest: people do not always behave strictly rationally, and an 'easy' attack may in some circumstances seem rational to the aggressor.

Each of the three armed services has an atmosphere and, to some extent, a personality type or types all its own. I rather expected to find the British Army hidebound and its controllers bone-headed. I could not have been more wrong. The Army has got very used to putting itself across to civilians, especially in the testing conditions of Northern Ireland, where flea-brains could do great damage. I expected the Navy to be bold and clear-sighted. On the bridges of ships, and in their operations rooms, the Navy *was*. But sometimes an awareness that the spotlight of public recognition and sympathy has fallen largely on the land and air forces (except in a few specifically Naval towns) left officers and ratings defensive about their job at a time when over 90 per cent of our essential supplies, including food, *still* come in by sea.

I expected the RAF, with fewer archaic traditions to contend with, to be straightforward and businesslike in explaining itself to a civilian; and it was. In that sense, there were fewer surprises for me than with the other two services, though just as many fascinating human insights into the recruitment, training and use of RAF men and women. Even in a hi-tech service like the RAF, where they are careful to keep the newest secret equipment well away from you (though you would not understand a thing about it if you examined it under a magnifying glass for half a day), human character is vitally important. It is the character and the lives of the men and women of the RAF that interested me, and which I have tried to portray here – as a gesture of respect to a service which appears to have kept pace with the times and fought its corner in Whitehall with some success for adequate (rather than wildly generous) funding, while spending quite a lot of its time and resources in peacetime doing useful things for the civilian community, like rescuing them from sickeningly high cliff faces (an activity I was allowed to share in a helicopter hovering perilously close to a Kent cliff four hundred feet above the sea), or flying them to hospital with understandable heart attacks from cross-Channel ferries crammed with holiday-makers.

I appreciate the help of the public relations department of the R A F in providing me with the facilities I asked for without obtrusively pushing their own official line or getting between me and the people I was interviewing. The conditions under which all those who talked to me for this book did so were the same as for my *Soldiering On* and *Ruling the Waves*: that individuals be freely quoted by rank but any names given should be fictitious in order to achieve the maximum frankness without embarrassing individuals; and that the finished book be submitted for correction on points of fact and security only – apart from which the ultimate editorial control remained in my independent civilian hands.

I am grateful to all those who gave me their time and candour. Though this finished portrait of the people of the present-day Royal Air Force is dedicated to them and their colleagues with respect, I have no means of knowing whether this will invariably be welcome, since I have tried to make this contemporary portrait a true one, warts and all.

It may inevitably be more sparse in picturesque historical oddities than those on the British Army and the Royal Navy that go to make up *The People of the Forces Trilogy*; but in the Royal Air Force, too, I found much human richness, much folklore and often a saving and healthy sense of humour.

PART I

THE FRONT LINES

I

The Home Defenders

'Every little blip on the screen which can't be easily identified, and we scramble for real. We even get called at this time of year because the Russians are going down to train at Cuba, where the weather is nice and warm. We have to be airborne within ten minutes of an alert.'

– Flight Lieutenant on fighter Quick Reaction Alert duty at RAF Wattisham.

'Fighting to defend the home of your loved ones is easier to rationalize to a young man than it is on a station where the intention is to go out and bomb some bad guys. A young man is easier to motivate, given that spur.'

– Station Commander, RAF Wattisham (a Group Captain).

These men are vigilant all round the clock and the calendar: the front-line defenders of the homeland itself.

Wearing full (and stifling) gear, two pilots and two navigators sit for twenty-four hours at a stretch inside a brick-built shed the size of a seaside bungalow, only feet away from their fighters. They have a standing instruction that they must be off the ground within ten minutes of an alert indicating a possible enemy air attack on the United Kingdom.

Unavoidably, they stink of body odour, in their thirty-pound gear, by the end of their day-long shifts at RAF Wattisham in Suffolk. During this time the centre of their testing lives is a black box smaller than a brick, called the telebrief, screwed on to the bungalow wall. This is wired direct into the station's radar site, and gives an up-and-down hooting sound whenever a Russian or other aircraft comes into the area surrounding Britain, usually through the Faroes–Shetland gap or the Iceland–Faroes gap.

RAF Wattisham is one of a web of four fighting stations in the United Kingdom which mounts Quick Reaction Alert systems round the clock. The others are at Leuchars near Dundee – the only QRA station in the north and always on top alert – and Binbrook and Coningsby, both in Lincolnshire, which share duty on Quick Reaction Alert on a rota basis with Wattisham, miles from anywhere and, reassuringly, almost invisible in the flat East Anglian countryside.

The Wattisham Quick Reaction Alert bungalow is built directly alongside a double hangar. Each hangar contains a jet fighter – Phantoms at the time of my visit, their prominent snouts and slender tailplanes suggesting housedogs sniffing for action. In the 'living room' of the bungalow, just big enough to contain a large television set and two small divan bunks, sat the four officers who would be the first to engage any aircraft that threatened the people of Britain itself.

The responsibility did not seem to weigh heavily on them. They were watching a Goldie Hawn video film and bargaining for an exchange of movies with a couple of officer joggers who happened to be passing. The transaction had to be conducted inside the bungalow itself, despite the overcrowding which began to suggest a Marx Brothers film. For the alert team to set foot even a few paces outside the bungalow door is a risk not worth taking for such a trivial purpose. When seconds count as much as they do when doctors correct a cardiac arrest in a hospital, frivolity is kept on a very short lead indeed.

Only one young man of the four-officer team (which is backed by a ground crew of half a dozen airmen) was *not* wearing his bulky and heavy immersion suit – that pair of padded overalls of porous plastic designed to enable the pilot or navigator to stay alive if he has to ditch himself in the North Sea or Atlantic up to perhaps seven hundred miles from home. This was because the navigator, a Squadron Leader, was, at thirty-two, the oldest and most experienced man there, and reckoned he could get into the hideously heavy but life-saving garment quickly enough if the alert sounded.

'Russian planes come over in support of Soviet naval exercises,' he told me, his finger pointing at a neat map showing the chief areas of concern. 'Sometimes the Norwegians have a look at them and then hand them over to us. Sometimes the Americans launch a plane out of Iceland to have a look at them. There is a certain amount of Soviet aircraft on their way to Cuba or Lusaka – I suppose for exercises there.'

He pointed out two deep trenches in the Atlantic, just to the west of the Shetlands, which are regularly used for Russian submarine exercises, since they afford sonic cover to submarines. Such Russian naval exercises attract air cover, and this in turn can trigger the Quick Reaction Alert system, which may send a plane out from Wattisham to look.

Sometimes there is only one scramble a week; during exercise time there may be many more. Sometimes a Russian aircraft is watched as it goes out on a long flight and then watched again later in the day as it returns.

'The Russian aircraft can stay out all day,' said the one Squadron Leader who was minus his immersion suit. 'They are into eighteen- and twenty-four-hour sortie lengths. They may come from the Murmansk area, go well into the Atlantic, south-west of Ireland, and turn round and come back in the same day. We have had people coming out to intercept at ten in the morning, landing again, and then going up again to see the same Russian aircraft coming home. Sometimes two aircraft have been launched.'

Two aircraft are always on the alert at Wattisham, but rarely are two launched. The second is there to provide continuous back-up to handle any new situation when the first fighter is airborne – or to be sent up if the first-in-line fighter cannot, for some reason, take off.

In the hangars, a civilian is struck by the immaculate hi-tech sheen of the actual aircraft and the comparatively primitive look of the machines which get the planes started up, the appearance of which gives the impression they could have come straight off a building site. There is a generator to supply the electrics while the plane is on the ground, and a compressor for pumping into the jets to promote the start-up explosion. Sometimes these fail – though there are, of course, back-ups; just as the electric button which opens up the huge hardened hangar doors has a back-up should it fail to move the enormous tonnage of the doors.

Getting aloft quickly has certainly become more complicated since the programme to harden (protect from bombs and missiles) as many parts of airfields as possible. Fighters used to stand on 'pans' – huge tarmac areas near the centre of the base – where they were easily accessible to hurrying crews. Unfortunately they were also accessible to an enemy. The decision was taken to protect aircraft in hangars that were proof against the largest conventional bomb – weighing one thousand pounds – dropping five metres (about fifteen feet) away, and also possibly proof against a direct hit by a smaller bomb. Coils of metal immersed in concrete provide a safe haven against blast and shrapnel, but they do not make it easier to get into defence fighters quickly.

Pilots and navigators, however, work out their own routines to overcome this difficulty. While the doors are opening after the pressing of the two-inch-diameter red button, an officer may struggle into the additional combination harness – a sort of Mae West, which gives him flares, a Pye locator beacon and a heliograph which will be of vital importance should he land in the sea. Wearing this, he will climb the metal ladder hanging from the forward cockpit of the aircraft and get into the cockpit. His helmet will be already there, with its electronics and oxygen supply connected: he merely has to put it on, wait for the thumbs-up sign from a member of the ground crew standing in front of the aircraft and taxi straight out on to the runway.

Often the aircrew will have to go on to what is called the Strapped-in State. This is where a plane has been seen around United Kingdom air space, but not been precisely identified. While radar and other means are checking the aircraft out, the number-one aircrew have to behave as if they are about to take off, but remaining in their cockpits inside the hangar.

If the crew does not have to take off after all, there is an inevitable anticlimax. This can be uncomfortable for young men who have to psyche themselves up for action for twenty-four hours at a stretch once every ten days, following it up with twenty hours free (but in practice often having to do other work). They will go back into the bungalow, and perhaps make themselves a solacing meal in the self-contained kitchen, or (another) cup of coffee. Or they will take off their Goonsuits (as the immersion suits are known) for the briefest of spells, perhaps leaving them near the door between the bungalow and the hangar so they can pick them up and put them on en route if they get another alert.

'We drink plenty – but no alcohol,' said a twenty-six-year-old pilot. 'We can smoke in here if we like, though as it happens none of us here does: it is not very popular on the base, and few of our squadron (No. 74) do. It is bad enough without that. You stink anyway at the end of this shift, and it gets very uncomfortable.'

But the pilot teamed up with the thirty-two-year-old veteran navigator, a Flight Lieutenant from a comprehensive school in the west country, dismissed the idea that the job could be a strain (except sometimes to the ears when listening to aircraft starting up on the ground). 'It is like every other job, the more you do it the better you get at it. It is like a bank manager or someone in industry; they work their seven or eight hours and they don't notice the strain because they are used to it. It is the same with us, the only difference being that we work for twenty-four hours at a time.'

Not the *only* difference, surely? How many civilians have to wait for an alert from a black box on the wall (or the station siren if that fails), say 'Scrambling!' and be into special clothes and airborne within ten minutes? What, I asked the veteran navigator, if the alert was for real – didn't the thought that it might be add to the strain?

I had said the wrong thing. 'We always *do* think the thing is for real,' admonished the navigator, 'because there *are* Soviet planes at the other end. In this peacetime environment it is very much a policing environment – that is the way it is written up in our instructions – but the adrenalin does get flowing nevertheless. The fact is, though, that you may spend four hours in the air to get to where the other guy is; and, by that time, you have calmed down, and it becomes quite routine. But it is *real* enough, what is happening.'

The larger reality can sometimes be soothed, if not blurred, by the ribbing on lesser issues that goes on freely between the Quick Reaction Alert crews. For instance, who is *really* more important, the pilot (as civilians might assume) or the navigator (as technicians might assume)? This little game was being played, off and on, while I was with the teams.

The veteran navigator said: 'The old Royal Flying Corps had the right idea. The officer was the observer and told the *NCO* at the controls where to take the aircraft.'

'You don't *have* to get on personally to make a good professional team,' riposted his pilot drily, 'but it makes things easier if you do. Seriously, it is more a fifty-fifty thing between pilots and navigators. You each have your own skills and you each have your separate job to do, but if you do them well together, it makes things easier. We are all aware of the importance of our job, and do it together.'

Such readiness for immediate defence of the homeland (for the fighters fly armed with missiles bearing live warheads), *and* the routine flying which surrounds it at Wattisham and the other Quick Reaction Alert bases, exacts a social price in the nearby civilian communities. The QRA men inevitably realize the importance of their job, but civilians – especially younger ones – may be pardoned for not being equally aware.

The little village of Bildeston, which is virtually directly under the flight path of planes taking off and landing at Wattisham four miles away, has an inn built in 1495 when Henry VIII, who certainly never had to fight an air war, was only four years old. It also has many Elizabethan timbered cottages, a square brick-built clock tower from which all but one of the four clocks have long since disappeared, a Co-op 'supermarket' about the size of a postage stamp – and sometimes an amount of aircraft noise that would outdo Heathrow international airport.

Were it not for Wattisham's presence just up the road, Bildeston's record for noise would have been established in 1855, when seven policemen had to quell a riot in the Crown Inn after the rioters had seized large quantities of ale and smashed all the windows and most of the glasses and bottles. The RAF, needless to say, were *not* responsible. Officers visiting Wattisham and boarding out at the Crown now have to reconcile the modern aircraft noise with the inn's much publicized ghosts – the gentleman in the tricornered hat who sits in the corner of the lounge bar (though not when I was there), a lady in grey who haunts the stables (ditto) and two children in Victorian dress who pop up all over the place (ditto). Perhaps during my stay they were all put off by the aircraft noise.

One antique dealer in the village waxed philosophical as he french-polished a nineteenth-century bureau. 'I am just old enough to remember the war, when you were *glad* to hear the aircraft noise because it meant someone was doing something for you. You can get used to anything, but the trouble is that young people today don't know what war is like, and take peace for granted.'

There was a double standard which offended him in the attitude of some of the public. 'People deplore aircraft noise, and then expect to go abroad for their holidays on aircraft. I say to them, "Do you realize that when you go on holiday, some of that noise you object to is yours?" I used to live fourteen miles away from Stansted airport and people used to say, "The noise is a disgrace." I told them, "You ought to consider yourself lucky the airport is so near to you when you want to use it – would you prefer to go all the way to Gatwick?" But the trouble today is that people take *most* things for granted – they have never heard of food rationing, have never had to make do with a tenth-hand bike as we did in the war – it's a new moped, motor cycle or car now. Therefore the reason the Wattisham planes are there is overlooked.'

Another shopkeeper said that most people said they could live with the noise, except when it became worse in the limited time of exercises. 'The RAF say they're going to upgrade the station and we don't quite know what that means . . .' (Presumably it meant hardening it up with reinforced buildings.) 'But I don't think the RAF will get away with *too* much, because we have some powerful voices living near here. But we must remember they have a job to do.'

Formal complaints to the station are regular, but few. They are lessened by the fact the local traders know that the RAF presence means extra trade. They are increased by the fact that in this area of Suffolk, some ten miles from the industrial town of Ipswich, there are a large number of cottages used as second homes. Prosperous Londoners or Ipswichians buy the cottages and use them at weekends only – when, by and large, flying at Wattisham (in peacetime) stops. But eventually the cottage owner retires, sells up in London and moves permanently to the 'restful quiet' of his country cottage, only to find that from Monday to Friday there is much more aircraft noise than he had previously encountered at weekends.

Sometimes protesters get blunt answers. One woman rang the station to say she was about to buy a bungalow near Wattisham and was disturbed by the amount of aircraft noise. She was told, politely but succinctly, that the station had been there since 1939, was likely to stay, and that she should ask for a few thousand off the price and use some of it to double-glaze the windows to blot out as much of the aircraft noise as possible. The officer who dealt with the call told me he didn't know whether she had bought the cottage on that advice; what he *did* know was that property prices were steep enough in the immediate area to persuade people like himself to commute at weekends to homes nearly fifty miles away. This did not suggest that aircraft noise had ruined the property market in the immediate vicinity of RAF Wattisham.

Placating the local populace is part of the role of the station's public relations staff. It is also part of the role of the Station Commander, who is also the key to keeping the mix of servicemen on the station both happy and efficient. The Station Commander – a man with an incisive face that seemed to be made for cutting through steel – assured me that the station as a whole believed in close links with the local community. In the past year, for instance, the station had raised £20,000 for local charities: 'National charities, I feel, get a lot of publicity and have ways, without us, of making a lot of money. It is the people around here who get less than they might. I aim to support them all I can with any spare money.'

The Station Commander, a pilot in his forties, had some of the problems of *any* RAF station (a unique sort of mixed community) and some problems special to Wattisham. There were two rival squadrons on the base when I visited it, 74 Squadron and 56 Squadron, both using Phantoms, but each with their individual history and traditions. 56 Squadron, The Firebirds, flew Hurricanes in the Battle of Britain, Typhoons from 1941, Swifts from 1954, then Hunters and Lightnings, until they adopted Phantoms in 1976. They have their own emblem, a Phoenix badge with red-and-white checks, which is on all their aircraft. 74 Squadron, The Tigers, had, as one of their top-scoring pilots in the First World War, Major Edward Mannock, VC; they were disbanded in 1919, reformed in 1935, were based on Malta during the Abyssinian crisis of the 1930s and made offensive sweeps of German-occupied France in 1941. They supported the D-Day invasion of the Continent as a unit of the Second Tactical Air Force. In 1960, with the introduction of the first Lightnings, the squadron became the first to be supersonic. The squadron was later made dormant, only to be reformed in 1984 to fly Phantoms, adorned with their distinctive yellow-and-black bar badge. 74 Squadron is certainly not afraid to overdo its connections with the tiger emblem: on my visit, I saw real tiger heads on walls, tigers in rugs with the heads still attached and more photographs of tigers than you would come across in a museum of natural history.

Such squadrons have a life of their own, as well as a life on the station they happen to be on at any given moment. Then there are the officers and airmen who are on the staff of the base itself, divided into three distinct wings – the Operations Wing, the Engineering Wing and the Administrative Wing, each with its own skills and priorities.

Fusing all these elements together is a task which does not face the commanding officer of a cohesive regiment in the British Army or a ship or shore base in the Royal Navy. The Station Commander at Wattisham commands 1,500 people, and leads another 1,500 civilian workers. There are fifty officers and one hundred and fifty airmen with each of the two squadrons. The Engineering Wing has some five hundred people, the Administrative Wing two hundred people and the Operations Wing eighty people.

There is no room here for a tentative style of leadership that may (or may not) work in civilian life. This is especially so in view of the fact that a continuous programme of hardening is going on, while at the same time a lot of officers and airmen are spending a great deal of their lives out of touch with the station, in homes of their own, miles away. It was rather different when the station was opened in 1939, when all personnel lived on the base, to live up to the motto of the station: *Supra Mare Supra Terramque* (Over Land and Over Sea), and when Blenheims took off on what was then only an unmetalled field to make the first bomb attack by the RAF on Germany (German battleships anchored in the Schillig Roads near Kiel).

'I have probably the smallest air defence fighter base,' said the crisp Station Commander. 'But this is an unusual station, in that I have got only a small number of quarters on base – 46 officers' quarters and 147 airmen's quarters. If you do your sums, you will see I am not housing many of my people. I have got six married quarters sites spread out as far as twenty-eight miles away. On these six sites I have 467 families. Some people do what we call "bean-steal". They have their own house elsewhere but choose to live in the mess during the week and commute home at weekends. That approach has altered the face of the RAF somewhat. Whereas we used to be tightly based, we are not quite so close-knit. That is not a criticism. But the fact is that I have to work harder to provide the leadership and esprit de corps that is necessary to produce the efficiency that I like. And not just what *I* like: it is what the government has every right to expect from its front-line forces. Motivation is everything in this day and age. Were I to have all my quarters around me on base, then it would be easier to create the community spirit on which esprit de corps can be built up faster. But I would claim that we are without doubt the most efficient and most operationally prepared fighter station in the United Kingdom.'

So, I imagined drily, would all the others? The Station Commander let fly at me with a list of records, awards and trophies till I began to realize what it must be like to be at the wrong end of a fighter missile.

Before the last tactical evaluation by a NATO team, the Station Commander called all of his 1,500 servicemen together in a hangar and gave them a speech that would have done credit to Noel Coward speaking as a naval officer from the bridge of a cinematic HMS *Kelly* as he took command of her to go to war (a speech said to be based on an actual one by Lord Louis Mountbatten).

'I told them what I knew,' said the Station Commander. 'That is, that they are bloody good. I told them to go into the evaluation and enjoy it, that they probably knew more about their jobs than the evaluators; and that provided they didn't suffer stage fright, they would prove themselves the most efficient station in Strike Command. The fact is that this station has thrived on success for the past two years.'

You had to know your people to run a station, said the Station Commander. Otherwise you didn't know how hard you could press them. He made station inspections once a week, talking to as many people as possible, and learned more this way than he did from the papers passing across his desk. He encouraged people from all walks of life on the station to play sports together, and played a lot of squash himself. He tried to set clear targets, so that everyone would know that if they failed to reach them they would get a good kick in the backside; but that if they met them, they would get praise and perhaps a day off. He had cancelled his last two inspections, because the people had done well and warranted a gesture of confidence. He believed in the open office door policy, and in working harder than anyone else on the station: he would do three or four hours of paperwork at home every evening, between six to eight hours at weekends. You had to have a clear idea of the weaknesses and strengths of the team of executives around you.

What did he consider his own strengths and weaknesses? 'Neither is for me to say. I would be happy if you asked any of the Wing Commanders under me what my weaknesses and strengths are. I know the things that make me hard to work for and what makes me easier to work for. But you ask *them*.' (I didn't, rather fearing that I would be mistaken for an agent provocateur or boss's nark; I chose to wait until later for the Station Commander's own assessment of himself.)

When I moved out to look at the two individual squadrons, I found these had their communication difficulties, being based on either side of the main runway and requiring a car journey of three miles to get from one side to the other, round the end of the runway. The route is described as Gasoline Alley, because it used to be a fuel dump; alongside it is a wrecked Canberra. This ancient aircraft periodically has holes put into it so they can be repaired for practice.

Each squadron has its own cookhouse and café, since a three-mile car drive from a far-off cookhouse could somewhat blunt a readiness for immediate air defence of the realm. Each squadron thrives on the supposition, expressed as a certainty, that it is the better of the two. This rivalry is encouraged, as long as it remains good-natured and a means of achieving efficiency rather than ruining it.

Most of 56 Squadron was away on other duties, but the acting commanding officer of 74 Squadron, a thickset Squadron Leader with a countryman's appearance, told me: 'We are quite lucky. There is always a certain amount of rivalry, which gives you something to aim for – to be better than the other squadron. But we have always helped one another, and we certainly have a good relationship with 56 Squadron. I have been on units where the rivalry has been such that people haven't helped one another. That is detrimental. That doesn't happen often.'

Some things that *can* happen can be poisonous. The squadron not in the front line at a given moment is expected to do other jobs on the base. Sometimes, although acknowledging it is their turn, a squadron will ask its sister squadron to supply one man to help with these duties for an afternoon. If relationships are bad, the request will be denied – on the philosophy that all's fair in love, war and competition. In such cases, the Station Commander may have to point out that the war envisaged is *not* one with another part of the RAF, but with an enemy.

Relationships between squadrons on bases like Wattisham have not been helped by the hardening programme, which has meant that aircraft from different squadrons no longer mingle on the large pan, but have their own hangars.

'It does give us a problem in getting aircrew to mingle,' said the Squadron Leader of 74 Squadron. 'To get from one squadron to the other will take five or ten minutes, depending on the traffic lights at the end of the runway. The fact that more people are living outside the station also creates a problem with intermingling. Some people arriving from off base enter it at the crash gates (the rumble strip like a cow-catcher in a narrow road on the base), bypassing the main domestic sites where they could come across other people. They don't see or meet anybody.'

The third factor creating problems for inter-squadron mingling, said the Squadron Leader, was the fact that house prices in the area were not cheap, partly owing to the fact that there were a large number of American bases in the vicinity, and American demand for houses forced up prices. With a heavy mortgage to pay off, money to spend on social events was scarce.

I was to get confirmation of this on the occasions I ate in the mess. Most officers said they regarded the regular dining-in evenings, when it was virtually compulsory to eat in the mess, as a nuisance rather than a pleasure. The bar was mostly deserted, and when I stood my round I found that, in addition to my own tomato juice, I was buying an orange juice, a pineapple juice, a half-pint of shandy and a single gin and tonic for an officer who had nothing vital to do for the rest of the day: rather a dent in the image of hard-drinking young men protecting the flag, I thought.

Work dominates people's lives. Each squadron is divided into two, A and B flights, each under a Flight Commander. They work different shifts, one in the morning, the other in the evening. It is no inducement to a night's heavy drinking if you know that at 7.30 the next morning you must meet in the tiny claustrophobic Meteorological Briefing Room for a weather view, that you will then be flying until three in the afternoon, when you will be debriefed, not going off duty until five in the afternoon or still later if you are on Quick Reaction Alert. The night shift comes on at 4 p.m. and lands at 11.15 p.m., meaning debriefing may overrun midnight – not a regime that makes alcoholic socializing possible, let alone attractive.

Life at squadron level is made even harder by the fact that some flying – 20 per cent – has to be at night for training purposes. In the light summer evenings, this means a late start in the day and probably a late finish.

In winter there are other problems which encourage a sober attitude to life. Sometimes the weather is so bad that flying has to stop altogether. Whether aircrew can go ahead in bad weather depends on their rated instrument skills. If they are rated at the lowest level of skill – called White – they must be able to see the runway from four hundred feet height before they are allowed to land: if they can't see the runway from this height they must overshoot and try again or make for the runway designated as a weather diversion because it is far enough away from the runway to have different weather; or even, in an emergency, the designated crash diversion, which is usually as close to the runway as possible, to make a quick touch-down possible.

The environment of squadron headquarters is not conducive to a flippant attitude to life. There is a 'soft' brick-built office block which would be immediately vacated in an emergency in favour of a hardened bunker with two-foot-thick reinforced-concrete walls, the heavy steel doors masked by a two-foot-thick blast wall. Inside this is an introduction to Armageddon – decontamination rooms, where clothes affected by nuclear radiation, chemical or biological warfare could be shed and treated; vapour hazard rooms where respirators hang as a reminder that even when clothes have been removed, potentially lethal vapours can remain on the skin; and an airlock room in which personnel must stand for about three minutes before they are free to enter the toxic-free area where professional life goes on more or less as normal.

The 'normality' includes three briefing rooms, one of which is used for delivering what are termed 'parish notes' – usually one-liners about the previous day's performance, especially what went very right and what went very wrong – and receiving complaints from junior officers.

'It is *not* a trade union, though,' said one senior officer hastily.

There is also a Flight Safety Officer's Room. The job may be done by a pilot or navigator as a secondary duty; it consists of reviewing the day's activities in terms of safety, and sending out a signal or telex to any squadron using similar aircraft, or to the RAF as a whole, if any generally applicable moral is revealed by any mishap.

A thirty-two-year-old Engineer Sergeant, who had left grammar school in Yorkshire with three A-levels, told me he was on an unaccompanied tour and lived in the sergeants' mess. As senior NCO he was the shift boss for the weapons trade, and had a busy time – mostly rearranging aircraft for role-changes. It meant changing weapon loads, changing ejection seats (which shoot a pilot a thousand feet into the air, so that, even if he is almost at ground level in a crashing plane, he will have enough height to allow his parachute to open) and similar jobs. His day, he said, started at 7.30 with reading the 'diary' left overnight from the previous shift, and went on until 5 p.m.; or started at 4.40 p.m. and could end at around 2 or 3 a.m. the next morning. He was lucky if he had half an hour to eat his main meal. He had seven aircraft to watch; and the chances of all seven producing no problems at all on any one day were 'non-existent – aircraft get older, which means you get more snags'.

But the Sergeant stressed that he enjoyed squadron life. He had intended to become a teacher but realized just in time he didn't like children.

Had he ever thought of becoming an officer, which many people in his position chose to do? 'Oh, *no* – no offence to officers, of course. Having reached the rank I am at, I have got enough to do to reach the rank of Flight Sergeant, which is as far as I can go as an NCO. I can't see the point of getting to the top of my tree only to go to the bottom of another tree to start all over again with less money. Having said that, there *are* a lot of people who get to a high rank as airmen and then switch to a commission. But not for me.'

In the airmen's hardened block, I met a young airman, a twenty-two-year-old, who was on seeing-off and seeing-in duty. This meant that he had to be there, and active, when planes took off and when they came back to their hardened aircraft shelters (H A S). This involved duties concerned with covers and locks of undercarriages; winches; two pins; safety pins and servicing. Wrapping-up operations after a flight could take between forty-five minutes and one and a half hours.

But, he said, it was not this arduous duty he minded; it was the 'gash' duties around the squadron he had to do, which could include sweeping up with a broom. 'Yes, it is these gash jobs, and the waiting around, I don't like. One minute you are working on planes, which is interesting, which is what you join the R A F for, and the next minute you are sweeping up as if you were a cleaner. I know these jobs have to be done, and I know Corporals have to get involved, too, but they are not popular. Otherwise, the only thing I would say is that I hope to get a posting away from Wattisham, because Ipswich is my home town. Though it means I can live at my parents' home, I would like to see life in another area.'

Such are the minor grumbles of life at squadron level. It is, however, a way of life supported on a station like Wattisham by the three specialist wings, who have their own complex problems to face, and whose loyalty to the station as such is inevitably more permanent than that of the squadrons, who may be moved elsewhere as and when required.

Of the three, the Operations Wing probably has the most glamour, being the most concerned with the management of flying aircraft. The Squadron Leader who was second in command of the wing, which embraces three squadrons (called Air Traffic Control, Simulator, and Operations Squadron) took me to the War Operations Centre, a hardened building which has a back-up alternative.

It was basically a concrete box, one end of it an entry control area (I had to give proof of identity and be issued with a pass, though I had already been vetted as I entered the station) and the other end devoted to the so-called Plant Room, containing a flight planning room, a communications centre (forbidden to all visitors), a computer and British Telecom room and a meteorological cell. I found there were identification posters of Russian aircraft on every wall. There were also two ladylike women in a glass-walled office, on their own.

'They are a typist and a clerical assistant,' said the Squadron Leader. 'They are civilians, and whether or not they would stay if we were hardened down, and locked up in conditions of actual conflict, I don't quite know. We would like to think they would stay.'

The nerve centre was the ground defence control centre. This was dominated by a V-shaped desk and in operation would have a commander at the point of the triangle, controlling airfield damage measures, passive and active defence measures, dealing with personnel injuries and deciding, in contact with the Army, what measures were necessary for the airfield's defence. In a room off this big one, repairs of communications and the disposal of unexploded bombs and ammunition would be master-minded. In the Air Operations Room, there was a huge wall chart with markers, showing the state of readiness of all the fighters stationed at Wattisham. This room is the link between the aircrew and Higher Authority. 'The only thing we are interested in, during war, is the readiness state,' said an officer.

This is not quite true. There is something else which interests the men and women in the Air Operations Room, in peace and war: the position of Russian ships in British ports nearby. There is a chart on the display board giving the numbers of Russian ships in these ports. When I looked at it, the chart showed one in Boston, three in King's Lynn, two in Felixstowe, none in Ipswich and none in Harwich.

'The fact we record their presence is nothing to do with the prospect of hostile action against them,' said an officer. 'It is merely a way of saying to our aircrews: "Be careful what you say on the radio, you could be overheard by these ships." There is nothing secret about the information. You can read what ships are in the ports every day in the *East Anglian Daily Times*.'

The Air Traffic Control Tower, or parts of it, was less claustrophobic. It was a smaller, rougher version of the sort of tower you would expect to find on any civil airport, with the Approach Controller sitting in a windowless darkened room below the windowed top of the tower. The Approach Controller was a bald fifty-five-year-old, whose skills were such that his retirement was being postponed to an indeterminate date in the future.

'I am a general dogsbody,' he told me.

Not so. He was, in fact, responsible for all outbound flights from Wattisham, and for the radar service operating in the lower air space, between one thousand and ten thousand feet. As part of the transit service, aircrew in the air will be told the appropriate pressure setting of their altimeters for the area they are in. This is an absolutely vital job, since a wrong setting could cause a pilot to level off one hundred feet under the ground instead of one hundred feet above it.

Aircraft are picked up by Approach Control when they are forty miles from the station. They are kept until they are ten miles away – which, at four hundred miles an hour, means when they are one and a half minutes distant. Then they pass to the visual control of the directing officer in the glazed tower-top.

'In a day, we may deal with about sixty sorties,' said the Approach Controller. 'We stay with them outbound for five minutes only, whereas, inbound, we are with them for ten to fifteen minutes.'

Unlike a civilian airport, Wattisham does not have secondary radar – the sort used to determine how high an aircraft is off the ground. Aircraft are shown only in plan, which shows how far they are from the base as the crow flies, but not how far they are above the ground. Theoretically it would be possible, if the height measuring pressure device had been set wrongly, for two aircraft to approach one another at the same height. Hence a visual control from the tower is highly necessary.

'It is a job with definite peaks and troughs,' said the Approach Controller. 'Unless we get caught up with training we are often waiting for something to happen. But when it does happen, it happens quickly and is over very quickly. Whatever is happening is happening at six or seven miles a minute. There is a short-term high pressure; it is all go. Heathrow is like running a regular bus service; we are like an irregular taxi service. Everything comes back from a different direction; things have to be totally flexible. We have made a study of small civil airports, used by amateur, inexperienced pilots. The pilots give out positions different from the ones they are actually in, and they say they are at heights they actually aren't. And they are a big menace. We guard against that, we bear in mind that every approach is different: the job is a state of mind. I have been doing it for sixteen years, and the only disadvantage is that we still have to do a fair bit of niff-naff (paperwork) as well.'

It was a time of stretched nerves in the air control tower itself when I climbed up into it. A Squadron Leader in one Phantom was practising for an aerial display he was due to give the following weekend in an air show in Kent (where he would not actually be able to land because the runway was too short, meaning he would have to take off from Wattisham in Suffolk, give the display in Kent and then return straight home again).

Some of the stretched nerves were mine. As I stood on the roof of the control tower, the plane dived at me from ten thousand feet, pulled out of the dive near ground level, shot straight up again like a champagne cork coming out of a bottle and looped the loop through a passing cloud while simultaneously doing a roll. It then flew past at low level, upside-down, corrected itself and repeated the whole performance.

The noise was such that I had to go back into the tower itself before I could talk to the Wing Commander Operations, the equivalent in service terms of the airport manager. 'It's a statutory requirement, because it is low flying, that our practice displays are supervised – normally by the Station Commander himself, but if he is not available, by me,' said the Wing Commander Operations. 'The Squadron Leader is an experienced solo acrobatic display pilot who does displays at all the air shows. The locals hate it, but once people understand what it is for, they stop complaining. It shows our skills to the public and, at shows, it raises money for charity.'

The Wing Commander Operations said the acrobatic ability, which takes such pilots to within five hundred feet of the ground, would have no value in actual war. But here, I thought, he was underselling himself; for the acrobatic ability must make it easier for pilots to stay alive in a real encounter.

The Local Controller – the man in immediate charge of the air control tower visual control – was a tall bespectacled Sergeant who had been doing this sort of job for several years. 'We have no electronic aids at all,' he said with pride rather than complaint. 'It is right between us and the aircraft and its officers. We are concerned about what happens in the immediate visual circle. We clear aircraft to take off and land, and control them as they do a visual circuit.'

Was it very testing work? 'We do an afternoon shift one day from 12.30 to 6.30, the next day a morning shift from 7.30 to 12.30 and then a night shift – in other words, twenty-four hours over two days. You never know how testing it is going to be. It can be pretty exciting. Things go wrong – perhaps a hole in the runway, so the aircraft are restricted to only part of the runway. Or they cannot line up in the taxi way to the runway. With the squadrons on either side of the runway, they will try to integrate their timings to begin with; but, as the day goes on, perhaps one side is delayed, and then you have a lot of aircraft going off at the same time.'

Knowing that the take-off and landing rate at Heathrow airport is one every ninety seconds, I asked how fast they could cope with aircraft at Wattisham.

If the weather were fine, said the Sergeant, they could take aircraft as soon as they came into the radar frequency, and land them immediately. But they were restricted to four aircraft in the visual control area at any one time, though they often made a low-level circuit, under the clouds, as they came in.

The Sergeant said he had medical problems which, together with his age, meant that he would be stuck in the same rank till the end of his service. But he was lucky to be stuck in such an interesting job, the most interesting one in his twenty-six years' service, and one he had always wanted to do. If he had been able to move into the radar side, he could have had promotion to Flight Sergeant and Warrant Officer. But he was not really frustrated, because his job was one of the best, and most *immediate*, in the tower.

Engineering Wing had its own rather different domain, based in hangars and electronics workshops that could have been a civilian factory but for the fact that the operatives were uniformed, and its departments were called 'flights'.

The Flight Commander of the Aircraft Servicing Flight, a crucial role, was a tall, thick-waisted Captain with a boyishly intense face and penetrating grey eyes. *Captain?* Since when did the RAF have a rank of Captain? It didn't, I was assured: the Captain was an American on an officer-exchange tour of duty.

He saw me in his office, which he had made very much his own, and compared his experience with the RAF with his experience in the United States Air Force and with the German Air Force. 'It has been interesting, learning a different system, a different way of trying to accomplish the same aim. The concept of engineering is somewhat different here.'

In the USAF, they did not have squadron engineers – engineers attached directly to the individual squadrons – as they had in the RAF. In the USAF, all engineers came under the Deputy Commander for Maintenance, not under the control of the flying squadron. The disadvantage of the British system was that engineers on the squadrons were working for two bosses, the Flying Operations Commander *and* the Officer Commanding Engineering. Under the American system, all the engineers came under the control of the Officer Commanding Engineering; and if planes were not available as they should be, the engineer was not torn between two bosses.

This was too much for the American's senior officer, the Officer Commanding Engineering Flight, a Wing Commander as British as Biggin Hill, a slight man with lively intelligent brown eyes and a quietly assertive manner. He had come into the American's office during the conversation, and interrupted promptly: 'In defence of our present system, the Engineer has a corporate identity with the whole squadron. When you have a centralized system like the Americans, you don't get that speed of reaction and support that you get with us. We tend only to centralize when working on the larger aircraft, like the Hercules, when the focus is more on the individual machine rather than the squadron to which it belongs.'

Back came the American, anxious to be fair. 'On second-line maintenance' (the equivalent of a six-thousand- as distinct from a three-thousand-mile service on a motor car – taking in more aspects) 'we in Britain normally take five to ten days, and do things which in American terms would be almost *third*-line servicing' (amounting in motorists' terms to a major overhaul).

There was something else the American Captain liked about life with the RAF. His rank made him first engineering officer. As such, he had power to discipline airmen which he would not have in the United States Air Force. He had done several orderly rooms, and meted out relatively minor punishments; whereas in the United States Air Force such cases would have to be referred to the Squadron Commander, who would be a Major or Lieutenant Colonel. The Captain thought it right that he, being in the firing line, should have direct disciplinary powers over Corporals and below. He had given seven days' restrictions and, on one occasion, had ordered the payment of £50 to make good some damage.

The four servicing teams under the Captain's control were working together in one hangar on as many planes. A bird in flight had got sucked into the air intake of the engine of one jet and, as well as being shredded itself, had bent the frame of the engine. Seldom can birds do as much damage as they can when they are sucked into jet engines. This repair – the second in the two years the Captain had been with the RAF – would take between three and four months. A Jet Provost trainer aircraft was also in the hangar, because a bird had been sucked in during take-off, one of the most potentially lethal mishaps that can afflict any aircraft. The station, I was told, had a bird-frightening team, especially set up to go round the vicinity of the runway, playing tapes of bird noises through loudspeakers. These would be bird distress calls near the runway itself, or mating calls well away from the runway, intended to draw live birds away from it to a safe distance.

Other teams were attending to regular servicing, which takes either nineteen, twenty-seven or thirty days, depending on the state of the aircraft and the number of hours it has flown.

In a different hangar of the Engineering Wing, the domain of the British Squadron Leader, engines and ground-support equipment were being repaired and maintained. They included ejection seats, rockets and even explosive equipment – though the latter is usually kept in specially designated areas, for which the wing must get a licence from headquarters, much as a civilian wanting a handgun would have to go to the police. This part of Engineering Wing is also responsible for the two hundred vehicles on the station, ranging from staff cars and vans up to the 40,000-gallon aircraft refuellers and the two tracked armoured vehicles, a Spartan and a Scimitar, used by the explosives ordnance team.

There were 150 people in the Electrical Engineering Squadron, working on simple tasks (simple to them rather than to a civilian observer), and complicated ones – including the Phantom radar system and autopilot systems, which anticipate air bumps and so assure a more stable ride. The work is often a matter of taking to pieces a box which can be quickly removed from an aircraft once something has gone wrong inside it. Most of the electronics on a jet fighter are boxed in this way; in the Engineering Wing electronic workshops, they are taken to pieces, different circuit boards are put in, or the whole box is discarded.

Men doing this sort of work are well paid. Some said they would be lucky to get as much for comparable work in civilian industry.

The Wing Commander in charge of the whole Engineering Wing told me that he earned a basic salary of about £18,000, plus a generous boarding-school allowance for his children. 'The total package would be difficult to beat outside the RAF,' he argued. 'Yes, people do leave the service and go into industry and reach quite substantial heights and gain substantial incomes, but this is usually later, when they have worked their way up. Initially they usually have to take a drop in salary.'

I wondered whether the electronics of the RAF would ever be included in any government's privatization plans – bearing in mind that naval dockyards, for instance, had been hived off in this way. The Wing Commander, who pointed out that, under the present system, some electronic equipment was already sent back to the manufacturers for very complex repairs, said he had never found that private companies, as compared with nationalized undertakings, were any poorer or any better. What he did find was that smaller organizations tended to provide the better support, because they were anxious to have the business. In this respect, at least, the RAF appears to be rather like any other customer.

A civilian might assume that the Administrative Wing was the tamest and most bureaucratic of the three, and that life was devoid of both initiative and excitement. In fact, since the wing involved conjuring up the right supplies when required, individual initiative is not at all out of place.

A Flight Lieutenant who was acting boss of the Supply Squadron of the wing when I visited told me: 'It takes, I wouldn't say brigandry, but a lot of initiative and background knowledge as to where to look, and how to go about it. The supply management branch, for instance, holds most of the critical supplies for the station. Now, there are quite a few stations with Phantoms. Obviously some spares are critical to *all* squadrons. If we happen to know someone at the supply management branch, then it is a lot easier to call the guy up and talk to him, using his first name, and cajole that particular spare for Wattisham. If our guy didn't know anybody and just said, "We need that bit of kit", he might not get the same attention. There is a lot of that going on. It is a matter of pride that we can get our aircraft serviced before any other station – and supply is what it is all about.'

Anglo-American co-operation has paradoxically accentuated supply problems at Wattisham. Some of the Phantoms – not all – were bought from the Americans at the time of the Falklands war. They have American engines, instead of the Rolls-Royce engines in the British Phantoms. They have, indeed, some *twenty thousand* parts which are different from the British model.

While the supply administrators are trying to sort all this out, they also have to deal with calls for more mundane wares, such as toilet rolls. While I was talking to the Flight Lieutenant in temporary charge, a phone call came through, demanding some more paintbrushes. The American Phantoms are even a different colour from the grey of the British: they are a greeny grey which is more aesthetically pleasing, is more invisible against the sky, but creates problems for the painters.

The department also looks after the married quarters, a duty which often resembles playing with a yo-yo. Some couples want to have their own furniture, so the RAF furniture is taken out and moved to the stores depot near Gloucester. Then another couple will move into the quarter without furniture, meaning it will all have to be brought back again. The RAF handles the furniture for all three services, under the system by which such mundane but necessary services are farmed out to one service only; but despite its experience, the supply of furniture is a topic almost fraught enough to deflect thoughts about a possible Third World War. Indeed, some officers and airmen joke that the vying for acceptable furniture *is* the Third World War.

At the headquarters of Administrative Wing, they look after the financial and welfare aspect of people's careers, a task also requiring the judgment of Solomon. There are five RAF men here who look after three hundred people each.

If an airman gets involved in a fight with some locals, this department appoints an officer to attend the civil court hearing and write a report on it – which will determine whether further action will take place at RAF level, since it could be argued that the man has brought the RAF into disrepute. People who do not dress properly will be faced with summary action. Those who go absent without leave are likely to be referred to the Station Commander. Two had been referred in the past few months, I was told; each got twenty-eight days' detention. But not every charge is heard, and no charge is made without first taking advice from the legal department.

A Flight Lieutenant in the department told me he sometimes had a problem of conflicting roles: one minute he might be a father-figure to a man, advising him about his career and finances; the next minute he might be reading out a charge against him. It was just as well, he said, that disciplinary problems were not particularly frequent in the RAF: RAF men were intelligent men using their specialist qualifications, unlike some services he could mention.

There was one further boast the department could make. Wattisham was the first station to have a computerized personnel management system, though other stations are now following suit, and the ultimate objective is a RAF fully linked up by computers, all of whom can talk to one another. At the moment, postal missives between stations via the Post Office can take three days, which is not thought appropriate to the twentieth century, let alone the twenty-first.

Morale on crucial stations like Wattisham will never, of course, be transmittable by computer. It rests on personal contact and that spirit of friendly competitiveness I saw expressed in a couple of memos pinned to a mess notice-board. The first, from the officers, extended an invitation to the 'members' of the sergeants' mess to play them at hockey and cricket. The reply underneath said that the '*Gentlemen*' of the sergeants' mess would be glad to play the officers and 'give them a thorough dicking'.

That sort of thing, said the Station Commander as I left Wattisham, could be more important than the heavy hand of discipline. Then and only then did he confide what he thought his strengths were as a Station Commander: he could delegate, he could give clear targets and he could regard the people under him as his most valuable commodity, to be carefully looked after. And the weaknesses? Well, he thought he had a 'short fuse' about some things, including mistakes made more than once. He could also give too much attention to detail, though once standards slipped in small ways they could also start to slip in big ones. And one *always* had to remember how vital the role of a station like Wattisham was to the defence of the United Kingdom.

Strangely, it was an airman working in the parachute shop who focused most vividly for me the crucial nature of almost all activity at Wattisham and RAF stations like it. He was stowing one of the parachutes used to help stop a landing jet – by being released from its tail. It was intricate work that baffled a civilian eye, and it took him about a quarter of an hour.

What about the parachutes that guarded the lives of pilots and navigators? I asked. Oh, said the airman, that was complicated and technical; he could just about repack two a *day*.

Pilots and navigators (pilots costing around £3 million to train) entrusting their lives to a diligent airman – it said much about the co-operation required on a local Air Force station that directly defends every man, woman and child in the country. The sweating men in the bungalow who wait for the Quick Reaction Alert need all the backing they can get.

Just as certainly requiring good backing are those men and women of the Royal Air Force who defend the homeland from overseas bases.

2

West Faces East

> 'None of us is here in West Germany casually to kill or be killed, but if we had to go to war to stop their invasion, we would ruin their whole day. I don't think anyone would say that within a fortnight we would drive through Moscow in a victory parade, but we would hold them far enough away and long enough to enable the politicians to do their job. We are defending Dover at Helmstedt, if you like.'
>
> – Squadron Leader, headquarters of RAF Germany at Rheindahlen.

> 'I don't think anyone seriously thinks Berlin itself could be defended. I know some officers who keep two bottles of vodka in their desks, on the theory that if the Russians come, we might be able to get them drunk.'
>
> – Flight Lieutenant, RAF Gatow, West Berlin.

At RAF Gütersloh, the nearest base to the East–West German border, there is a small oak-beamed room at the top of a square corner tower. In it, fifty years ago, Hitler's deputy Hermann Goering liked to tell young Luftwaffe pilots tall stories about his own flying days.

Some of the stories told by the misleadingly roly-poly Reich Marshal, an air ace in the First World War, were *so* tall that on one occasion, after the umpteenth beer, the more valiant young pilots were emboldened to question their absolute connection with the truth.

'If what I say is not true, may that beam drop on my head,' retorted the Reich Marshal, pointing to the centre beam running directly over the table and his chair.

The young pilots were to hear this expression rather frequently. So for Goering's next visit, they doctored the beam, splitting it on a hinge and running a wire from it to a tiny trapdoor in the floor, at an inconspicuous corner of the room.

The inevitable moment came when the Reich Marshal sought to prove his veracity by challenging the beam to fall on his head. This time, with a clatter, the beam promptly sagged in the middle towards his sleek scalp – to the jovial delight of the second in command of one of the most evil regimes the world has known.

This uneasy blend of grisly reality with the apparently cosy and human could, fifty years later, stand as a metaphor for the tone of life experienced today by the 2,500 serving officers and other ranks and the 3,000 relatives and civilian workers at RAF Gütersloh, the only RAF base east of the Rhine, and the busiest of all the bases in West Germany (which, between them, have over 10,000 RAF personnel and double that number of relatives and civilian workers).

On the one hand, Gütersloh is a small town in Northern Germany, with neat rows of vine-covered homes, mothers driving their children to school in the family's second car and Goering's tower, with its luscious ivy and shades of art nouveau, a tourist attraction for all those who live in or visit the little community.

On the other hand, as even the young boys and girls rushing about with squash rackets through their satchel-tops must be aware, the gracious charms of Gütersloh are just nine minutes' flying time away from the obscene walls, wire, land-mines, ferocious dogs and scatter guns of the East–West German border. Behind this depressing and squalid line, built by the Eastern bloc and straggling almost a thousand miles through Germany, the Warsaw Pact countries have assembled – whether they intend to use them or not – well above the three-to-one advantage in tanks, guns, missiles and men thought necessary to give a reasonable chance of success in invading the West; certainly many times more than they would need for merely defensive purposes.

The threat is understood by the personnel of all four flying stations of Royal Air Force Germany – the 3,500 at Brüggen, a hundred miles back, the 3,500 at Laarbruch, back further still, and the 2,000 at Wildenrath; but it may be felt more *personally* by those exposed at Gütersloh, the only station outside the cluster formed by the others.

'One of the things one becomes very aware of here,' the Wing Commander in charge of administration told me after I had experimented eerily with Goering's beam, 'is that *everyone* on the station gets a great deal of satisfaction out of seeing that everything works. We are unique here, and the reason the support side works better than anywhere else is that all our people know they are automatically integrated with the front line.'

The perceived threat would be on the same sort of lines originated by Hitler and Goering: the art of Blitzkrieg. The Warsaw Pact countries would seek to disorientate the defence by a mighty bombardment by ground and air, followed within the hour by a first wave of tanks and infantry, in such enormous numbers that they would penetrate deep into West Germany before anything effective could be done to stop them. This first line, regarded as expendable, would be followed by a second line before the defences could pull themselves together, to be followed by a third line and even more if required.

It is the precise role of the RAF in Germany to make sure this strategy is disrupted by blunting the edge of each successive advance – 'until the politicians can sort things out', as one officer put it while we looked over RAF Gütersloh's front-line contribution to that process.

It took the form of squadrons of heavy Chinook helicopters, lighter Puma helicopters and, above all, the squadrons of quick-reaction Harriers, the unique vertical-take-off aircraft that would move out into camouflaged sites in the field in the event of war, from there repeatedly to harass the enemy at short range. In a recent exercise, one squadron of fifteen Harriers flew several hundred sorties against the 'enemy' in seven days, a number of attacks that could not be rivalled by any other aircraft.

Complacency would be out of place. The peacetime Harrier homes, enormous huts like lakeside bandstands, are 'hardened' with concrete and steel mesh against all but a direct hit with a thousand-pound bomb; though it is estimated the aircraft would be already out, in their concealed and dispersed sites, long before such a bomb dropped.

But the problem here, as so often with things British in the present day, is simply money. There were not enough hardened shelters at the time of my visit.

The Station Commander, a short, energetic and ferociously moustached Group Captain, did not seek to minimize the difficulties. 'There is a small problem. Due to the number of aircraft and despite the hazards, it is necessary to put three Harriers in one shelter. They are almost wing to wing, with the attendant risks of damage. It is difficult, more difficult than we would wish. But the posture is to go out into the field in war, operating from over twenty locations. The Harriers might stay there for six sorties.'

Lack of money, and of space for expansion, cramps the whole of Gütersloh, not merely the Harriers. The Station Commander pointed out that the station had been built by the Nazis in 1937 for a population of only one thousand, compared with the 3,500 serving personnel today on the station whose motto is Ramparts of the West. In 1946, when the RAF took it over, some of the bombed hangars were repaired. Many are still standing today. It is impossible to extend the station westwards because of a canal and to the east because of Gütersloh town, which has extended itself almost to the end of the runway. One of the results is that, as the Luftwaffe had not provided any married quarters in the run-up to the Second World War (wives would have been in the way), the RAF's married quarters have had to be spread over thirty-four separate locations off the station, some of them as much as eighteen kilometres away – a drive of thirty-five minutes even when the roads are not busy. It is obviously not the best basis on which to fight a war.

Gütersloh has a working population the size of RAF Brüggen's to the south-west, but the station is only half the size. There are 170,000 air movements to deal with in the course of a year, and over 4,000 visiting aircraft.

'We were able to cope with 130 people arriving on an ordinary-sized aircraft, but today we can get jets carrying 300, and the organization has difficulties coping with that,' admitted the Station Commander.

While RAF men and their families sometimes have to live in rented attic accommodation in Gütersloh town, or special flats built by the assiduous Germans for rent, the helicopters are hardly more fortunate. Some of the Pumas have to be packed into old hangars put up by the Luftwaffe, where, unlike the families of the married personnel, they are not even out of the way if the station were to be suddenly attacked.

In the midst of such congestion, it is perhaps just as well that the Harriers would be sent off base in war, and that they spend quite a lot of time exercising away from base in peacetime – a process necessary if they, and the whole base, are to be fully ready for war well within twelve hours.

The Group Captain showed me a photograph of a dispersed Harrier, parked under netting beside a German house in the countryside. 'In wartime it would be *in* the house,' he pointed out. The house would virtually have been gutted and the Harrier pushed inside a hole made in the walls.

The men of 4 Squadron practise three times a year on dispersal sites. These are not those that would be occupied in war, which are never revealed or even hinted at by the presence of any training. The Commander of 4 Squadron told me: 'We don't have the amount of real estate we would like to have. Therefore we have to make do with what we can get. That gives us peacetime artificialities. We will use a green field. Though the aircraft is capable of landing on grass, we will put down metal planking for a take-off. That is one peacetime restraint. Another is that we are not allowed to damage trees, so our aircraft hides will stand proud of trees – whereas in war we could cut trees. That makes a site less realistic than we would like when we practise for two weeks, three times a year.'

At least, I was told, the Harrier crews do not have *one* problem, in peace or in war: fuel. Harrier pilots could take up their aircraft from their hidden positions as near the enemy as possible, do a rapid sortie against tanks and infantry threatening NATO forces, and then return to the hide, probably with a far-from-empty tank. So short are the envisaged sortie times that this is thought feasible.

But getting pilots to the state of practice where they could do this efficiently is a long-winded business. First the raw pilot has his basic training in computerized controls and weaponry, then he is put into a unit where his training is 'converted' into use specifically in Harriers. When he arrives at Gütersloh, he is already a pilot, but not yet an operational one. 4 Squadron or its sister squadron would take the man through air combat, low-level tactical flying, 'delivering' weaponry and doing reconnaissance.

'I will then declare him combat-ready,' said the Squadron Commander, a wiry man of thirty-eight whose ginger moustache was his only aggressive detail. 'It takes a fair amount of our effort. We get, probably, one or two new pilots every three months in the squadron, and the combat-ready work takes three to six months.' (It was almost with relief that I heard, elsewhere in RAF Germany, that there are *sometimes* more pilots than available aircraft.)

Even when they are combat-ready, the pilots spend a great deal of time exercising. Perhaps to the relief of local civilians, they will abandon Gütersloh for two or three weeks to go to Sardinia for weapons training on the air combat range there. It is run by the Americans for fully instrumented combat in, as near as possible, war conditions. When I visited Gütersloh, the squadron had just come back from its first exercise in firing missiles at towed airborne targets. Each year they spend two weeks doing low-flying exercises away from Gütersloh, when they can fly at their operational minimum of 100 feet as compared with the unrealistic 250 feet they are obliged to adopt in Northern Germany.

Pilots say that flying at 100 feet is very, very hard work. 'You have to work harder at avoiding the ground,' said one pilot. 'At 500 feet, you don't have to worry too much. At 250 feet, you start thinking about electricity pylons. At 100 feet, trees become a problem. You have to work harder at not killing yourself.'

It is not surprising that flying a Harrier, unlike riding a bike, is not something you re-adapt to easily after a period of absence. Pilots say that at the end of three weeks' leave they are not as good pilots as they were before it. 'You have to practise like a concert pianist, all the time,' said the Commander. 'After leave, you will not be as sharp as you were, and you will have to bring yourself back into form. You can still *fly* the aircraft but you have to relearn how to *operate* it.'

I said that I supposed it was hard on the local population as well as the pilots. Were there tensions with the local inhabitants? 'Yes, yes, yes,' said the Commander without hesitation. 'Noise is a constant problem. You don't fly over towns. You do your best not to overfly villages. You only low-fly where you need to do it in order to get some value out of it. It is becoming an increasing problem, as we get further and further away from World War Two. People are more reluctant to accept the reasons for low flying. And environmental considerations are more prominent than they were ten years ago. My house is one hundred metres from the end of the runway, and I know what aircraft noise is like. But we have to fly. Personally, I start work at eight in the morning and work right through until probably about six, without stopping for lunch. Probably we do about two sorties a day, at fifty minutes' actual flying time each.'

Faced with increased grumbles about noise from a German population too young for war to have any remembered reality, the RAF in Germany as a whole has staged an increasing counter-attack. It always emphasizes the positive benefits of having the RAF there, not merely in terms of defence (an unreal concept to some young Germans) but in terms of monetary self-interest (a matter more in tune with the times). The amount of trade the RAF brings to Germany is considerable – probably between eighty and one hundred million marks (over £20 million) to the Gütersloh area, and about the same for the other stations.

In the three messes of Gütersloh itself, consumption is high, and lucrative for the suppliers. Just short of one million meals a year are served, using 143 tons of meat, nearly five and a half tons of cheese, about the same quantity of butter, twelve and a half tons of sugar and about 28,000 litres of milk. (This compares with the 400 tons of food and 22,400 gallons of milk supplied by RAF sources to RAF Gatow in Berlin.) It is a very big-money operation.

But as far as Gütersloh's Harrier pilots are concerned, money is probably not the most tactful subject to raise. The Harrier itself, though not the most expensive of aircraft, has been in a sense slighted for financial reasons, among others. The Royal Navy is largely wedded to the doctrine of aircraft carriers (though described as through-deck cruisers for reasons of political prudence), and the United States Air Force is wedded to the doctrine of airfields. Both doctrines, on non-financial grounds, lead to the concept of the vertical-short-take-off Harrier being of secondary interest. The United States Marines, the Indian and the Spanish air forces have Harriers, but other nations who might want them have said they can't afford them: they cost about £10 million, and their pilots are too expensive to train.

It is, however, in terms of *personal* finance that RAF men and their wives at Gütersloh and other bases experience the most discomfort. Germany used to be a prized posting. It still is, but the days of wine and roses are definitely over, with the Deutschmark standing at around three to four to the pound sterling compared with the ten or even twelve of a couple of decades ago.

But the Wing Commander commanding 4 Squadron did not mention personal money matters when I asked him what he felt were the peaks and pits of his job in the late 1980s. Instead he put his finger on a problem I had heard of before (and was due to hear of again) in the RAF at home and abroad.

'The peak is the flying, and the pits is the paperwork,' he told me. 'The paperwork increases the higher the rank you go. This is my last operational flying tour of duty; my next will be as a Station Commander. Therefore I intend to get in as much flying as possible, and make the paperwork take its place behind that. Paperwork has inevitably become more extensive, or more important, because we are increasingly coming under financial pressures to do more with less. You are having to make better use of your resources to save money in the financial budget, so as to allow you to get extra equipment. You are constantly justifying yourself – more than in the past – which means more paperwork. If you don't get your paperwork right, next year you will find you can't get the money to do something because you didn't justify it this year. Paperwork is likely to go on becoming more and more important in this sense, I'm afraid.'

On the other side of the runway at Gütersloh, dramatizing a slightly different perspective on life, I found the huge Chinook helicopters under the control of 18 Squadron. The Chinook helps to transport men and heavy equipment, both for the RAF itself and for the British Army or other NATO armies. Its twin rotors are prevented from smashing themselves to pieces by huge synchronized gearboxes which periodically have to be changed, and its interior is almost as versatile (and stark) as that of a Hercules trooper. Both can be adapted for people or freight; and the Chinook can, and does, carry almost anything to and from Gütersloh – even a light tank or a Rapier surface-to-air-missile dispenser.

But arguably the Chinook, which may do most of its work supporting the British Army of the Rhine at one hundred miles an hour, while the Harrier goes six times faster, is the less glamorous part of the partnership. I put this possibility to a Chinook pilot of 18 Squadron (a twenty-six-year-old Flight Lieutenant) as I lunched with him and members of his ground team who work closely with him on the squadron, owing their individual loyalty to *it* rather than to an impersonal servicing organization.

I thought it was going to ruin the lunch. 'We are *not* the second division,' he said competitively. 'Basically, a helicopter is something that doesn't want to fly. It's like a bumble bee, which scientists say can't possibly fly, though it does. Even with the automatic pilot, which is very sophisticated now, you work hard to fly the machine on course. It takes a different co-ordination to fly, as compared with a fixed-wing aircraft, I can tell you! We have to fly in the dark with eighty troops, and dangling something from three nets below. It is quite a job.'

The Flight Lieutenant pointed out that the Chinooks had to work closely with the Army much of the time, and often had to share the same sort of conditions the Army had to put up with in the field as they moved them about. 'It is rougher than life with Harriers.' They lived under another tension: at Gütersloh they were well in range of enemy missiles. They had to be able to get away from base very quickly, and for that reason they constantly exercised in moving away quickly: 'Everything is always ready to go out straight away.'

In these circumstances, amiable relations with ground crew are essential. Often, it appears, they go well beyond the call of duty. A twenty-eight-year-old air safety equipment worker, of Indian extraction although born in Leicester, said that wherever the aircrew went in the field *they* went too. 'It makes life more interesting because you are working closely with the aircrew all the time. We look after what the aircrew wear, suits, helmets, jerkins, survival equipment, respirators. Even repairs of punctures to tyres of their cars or bikes. Or we let them have glue, so they don't have to go to specialist shops.'

'Yes, they do us favours,' chipped in the pilot. 'Sew our badges on; things you can't always get done elsewhere. We always keep friendly with them.'

I asked the air safety equipment worker whether his trade could draw the line on things they *wouldn't* do to further this spirit of accord. He had to rack his brains to think of an instance where his team hadn't been able to oblige on the off-duty front. Then he remembered the keen motor cyclist who had tried to get a microphone system installed in both his crash helmet and that of his wife, who habitually rode pillion with him. 'It didn't work. By the time we cut holes to fix the microphones in, the helmets were so weakened they were nearly falling to pieces. But normally the only time we say, "Sorry!" is when we can plead we are too busy.'

One of the reasons such crews are glad to work closely with one another all the time is that they all feel a little 'out on a limb' from the personnel of the other clustered RAF stations some hundred miles further back from Gütersloh and the front line. But there are men and women at Gütersloh who feel that, paradoxically, they are *closer* to a broad range of people from the West.

I found the Visiting Aircraft Servicing Flight in a hangar just off the 'pan' where foreign aircraft stood: a Danish aircraft, two French Mirages, a Dutch Army helicopter, a British Army Alouette helicopter, a Jaguar, a VC10 and an Andover.

'This is one of the busiest servicing flights in the RAF,' said the Flight Lieutenant who was its officer commanding, a tall twenty-eight-year-old with a cropped hairstyle and a Belfast accent. 'We deal with four hundred aircraft a year, and with four hundred air movements (arrivals or departures) a month. I am servicing about one hundred different types of aircraft a year – we have limited depth, but we specialize in breadth.'

I asked the Flight Lieutenant to give me some idea of what his normal working week might amount to. He obliged.

Monday and Tuesday: Involved in station exercise simulating an emergency.

Wednesday: Retrieving and towing away a Phantom aircraft whose brakes had failed and which therefore overshot the runway.

Thursday: An even more dramatic incident with a Dutch NF5 fighter, a twin-engined jet, which missed the wire across the runway with its arrester hook, and failed to slow down sufficiently until the hook caught on a barrier at the end of the runway, spinning the aircraft round off the runway and bursting the port tyre. The tyre had started to come off the rim, so to help get the aircraft moving on tow, an RAF Corporal of the Visiting Aircraft Servicing Flight gave it a sound kick, unfortunately getting his foot caught between the wheel rim and the tyre in the process. Diplomatic relations were smoothed only when the aircraft was reversed sufficiently to release the Corporal's foot, and was then towed off to the aircraft hospital in the flight's hangar while the Corporal was taken to a local hospital.

Friday: A Harrier caught one of its undercarriage legs on the approach barrier of the runway. It bounced from wing tip to wing tip for about 2,000 yards of the runway, before slewing off it, out of control. A civilian crane had to be despatched to retrieve it.

Saturday: Saw off a Hercules trooper and went to the station's summer ball.

Sunday: Station disaster exercise with fifty simulated casualties. 'That sort of thing generates a lot of paperwork.'

Quite. The Logistics Squadron sometimes has similar problems, also sometimes with diplomatic undertones. This is the squadron responsible for fuelling aircraft, using cylindrical rubber drums that hold eight hundred gallons or enormous bags, in rigid frames, that hold 10,000 gallons. Much of the refuelling has to be done in the field, which means the German countryside; and spillages are not unknown. There are German farmers who robustly maintain, in such instances, that a field thus infected will not be able to grow good crops again for seven years – though it sometimes seems to the RAF that the land in question was not particularly good for crops anyway.

'I find German relationships increasingly more difficult,' I was told by a gnarled veteran Warrant Officer of the squadron. 'In the first place, because of noise, we can't fly between 12.30 and 1.30 p.m., because the Germans are at lunch, nor after seven in the evening. Now we have the soil environment to worry about, too. The water table is very high here, which means that a seepage of fuel can quickly get to the water under the surface and pollute it. I don't blame the Germans one little bit for minding. One gallon of aviation fuel can ruin one million gallons of water. The United Kingdom is way behind Germany in these environmental things. It may be unfortunate from a military point of view, but it is something we have to live with.'

Such matters may or may not affect the morale of the average RAF officer and airman at Gütersloh. As far as airmen and their wives are concerned, it is usually personal matters that create more difficulties. A Sergeant with seventeen years' experience working in the Personnel Services Flight made this clear enough. The RAF has some strict rules which can cause friction at times of personal misfortune. If a young airman loses his father, it is quite natural that his wife may want to go with him to the funeral in the United Kingdom. But the RAF rules provide only for the bereaved airman (or airwoman) to fly home for the funeral. It is true that the wife may slip into an 'indulgence' seat on the plane if no one wants it for duty reasons, and in this way she may have to pay only £9 or so to join her husband; but the uncertainty, and the very fact of being a supplicant, can grate at times of emotional stress. 'Most of the other rules can be bent a trifle,' said the Sergeant. 'But once we talk about money, it gets more difficult.'

Before I left Gütersloh, I went to see an example of that trade which tends to take the brunt if there are underlying tensions: a cook. Aged twenty-seven, and used to providing between forty and one hundred lunches daily in the selfsame officers' mess that was once the stamping ground of Hermann Goering, the quiet-voiced corporal inadvertently threw some light on why tensions can sometimes build up in sunny little villages like RAF Gütersloh, even in peacetime: sometimes even more so than when exercising in the uncomfortable conditions of the exercise field.

'In the field for exercises, it is extremely long hours. You are lucky if you get eight hours' sleep, and you are working all the rest of the time. But you tend to be *thanked* more for what you do when you are in the field,' said the cook.

Just so. In the human sense, it may be the very sensation of being eternally in the front line that makes things work comparatively smoothly at Gütersloh – and at other stations which, though geographically further from the East–West front line, can, in other ways, seem even closer to it.

Berlin, of course, stands on its own as an exception to virtually every rule. RAF Gatow in West Berlin is not *on* the front line. It is *behind* the lines of the potential enemy. It is encircled by East Germany; its runway is right up against East Berlin; it is over a hundred miles behind the Iron Curtain and is accessible only through one autobahn, a rail route and three air routes, one from the north, one from the south and one route situated between them.

Of the 10,000-plus RAF people in Germany as a whole, only 700 or so are in West Berlin – some 65 officers and 650 airmen. About a third of them at any one time will be working a shift system that allows them little leisure during daylight hours, which is probably just as well. Once you *know* you are encircled, it is difficult not to *feel* you are encircled – though the West Berlin routine has gone on peacefully for around half a century, with only one interruption. That was in the early days when the Russians blockaded the land route to West Berlin. The RAF and the United States Air Force responded by airlifting personnel and supplies into the city in a way intended to show the massive determination of the West. The Russians eventually climbed down, and have tried not to embarrass the other three occupying forces – British, French and United States – since.

In this uneasy but routine stability, British forces have differing roles. The British Army drives around West Berlin showing the flag. It does so in East Berlin, too, on reciprocal visits supposed to represent the accord that bound the Allies when Berlin was entered – an accord long ago breached in reality, but still there on paper. In contrast, the few hundred RAF personnel are there simply to run the military airhead at RAF Gatow. It is as if the British Airports Authority and British Airways had set up a separate division to run a top-security airfield with civilian workers. The RAF officers and junior ranks know perfectly well that it is highly unlikely they will ever have to fire a shot in anger: in any invasion by the East of West Germany, the likelihood is that Berlin would simply be ignored. Why take prisoner, on paper, those who are virtually prisoners anyway if the surrounding countryside is hostile?

In some respects life in West Berlin, for the RAF, is lotus land. The big department stores are alluringly there in force – not only C & A, but also Haufhof and Quelle. It is true that families may not be able to get to them because of the shift system. One officer told me that in two years in the city, he had not been able to do any Saturday shopping once. But there are attractions not restricted by the opening hours, like the beautiful surrounding countryside in what is technically Berlin. The rural village of Lübars in the French sector is an often visited attraction, giving a good sane sight of horses and cows: it is rather as if Chiswick suddenly petered out into green fields.

Even in the built-up areas held by the three friendly powers there are ways of remaining pleasantly sane. The Berlin Wall Museum may or may not be one of them, with its reminder of the barbed wire and sentries of the Berlin Wall, where visiting tourists are meticulously photographed by East German troops in tall watch-towers, a ritual show of paranoia that clearly demonstrates to any Westerner the nature of the system he faces across the wall. The Charlottenburg Palace has several centuries of survival to put the visitor in a better frame of mind. The Economat, a sort of French Naafi, is there to provide not only the French with provisions but the British and Americans with an instant aura of France.

At the Deutsche Opera traditional operas are interspersed with new ones such as *The Sinking of the Titanic*. At a new development, called Summit House, there is a cinema called the Jerboa, which has the latest releases. There is another cinema for those officers and men who may be bound to the Gatow airfield – the Astra, which stands right in the middle of the quarters.

Sometimes – though more rarely of late – the entertainment is free. Gatow is at the south-west corner of the city, and the airport borders the wall itself. Attempts to get over the wall from the East, by light aircraft or glider, are sometimes made. The practical problems of removing the aircraft from the runway, and the diplomatic problems that follow, have enlivened many a day.

RAF officers in West Berlin readily concede that their duties in Berlin may be easier than the Army's. The Army has three battalions in the area, but because they are alongside a big city, their training areas are limited – far more than they would be in the United Kingdom or other parts of the world such as Hong Kong, Belize or even West Germany itself.

'The Army needs more ground space for its training, whereas the RAF has no combat aircraft in West Berlin at all,' pointed out one officer.

But there are two clouds over what might seem, on the whole, to be a cushy life for the RAF man and his wife. The first is that, according to the briefing given to all personnel before they enter Berlin, the place is crawling with Russian and East European agents. An injudicious conversation in a bar may be back to Moscow by first post. Attempts may be made to compromise and use an RAF man in a sensitive job, especially if he hasn't got a wife with him. It may seem odd to have wives delivered to a location encircled by Warsaw Pact forces, but one of the benefits is that men are less ready to go astray. Berlin is one of the few postings where the officer or airman finds himself located inside a city – he is usually physically well away from the temptations of towns. The airman can easily find temptation in Berlin: for a start, there is a British bus service which makes getting about the city easier for anyone who might be too timid to argue the toss with German bus conductors.

'And Berlin,' one officer told me, 'offers *everything.* Yes, you can hear the fabulous Berlin Philharmonic Orchestra under Karajan or his successors. Yes, there are beautiful art galleries and museums. But any perversion can also be taken care of. This has obvious security implications.'

The second cloud over the RAF man's life – and especially over his wife's and his family's – is that if he wants to go away for the weekend, he has three hours of driving through East Germany to Helmstedt before he can *start* it; he cannot simply have a picnic in East Germany. If he is in a sensitive job, he will not be allowed to join other RAF chums on visits to East Berlin, either.

The effect of these twin clouds is usually suppressed at a conscious level: none of the RAF people I talked to confessed to any feeling of claustrophobia in Berlin. The nearest I got to it was: 'Because you have a chance to see Soviet soldiers if you want to, there is still a *Third Man* type of atmosphere about the place, if your imagination wants it that way.'

At an *un*conscious level, the sense of encirclement would appear sometimes to act on the imagination in veiled and indirect ways. Like seeing ghosts.

Ghosts? West Berlin, and especially RAF Gatow itself, is rich in ghost lore. If it is not to be taken literally, it surely must illustrate *something* – possibly that imaginations go astray in one direction when they are rigidly disciplined in another.

It is said that birds will not make their nests in some parts of RAF Gatow, because of supernatural presences. This is a mild example. There are stories of apparitions dating from the invasion of Berlin by the Russians in 1945. One of the girls who worked in the officers' mess at Gatow, a robust and sane girl, would do any job except one. She would not be first in the morning to go into the billiard room to pull the curtains and flood the place with light. She would give no explanation for this inhibition, she would not say if she had ever seen or heard anything of a supernatural kind, and she was tactfully not pressed on the point.

Another story going the rounds is about a young airman who took a picture out of his quarter window. When the film was developed, inside a sentry box that happened to be in the shot was a Russian soldier wearing a 1945 uniform. It appears that this supernatural photograph, which must have a considerable commercial value, has never been publicly produced, which argues an unusually strong spirit of self-denial in the photographer.

The important point is not whether the story is true or not: the point is that the story circulates. Imaginations are self-prodded, as with an aching tooth. People come to believe in *atmospheres*, even when nothing is actually happening.

One Flight Lieutenant said his wife would not go along the basement corridor in the officers' mess building to take her dirty clothes to the communal washroom at the end of it: there was an unnatural cold about the place. Her husband, reflecting that basement corridors do tend to be cold, started to make the journey for her one fine Saturday afternoon, when he was overtaken by a feeling that something was wrong. He ran as fast as he could to the washroom, presumably in the belief that evil spirits are not quick on their feet, and returned by another circuitous route through public rooms, despite his having to carry piles of washing. Nothing actually happened; but his imagination was working overtime on the question of atmosphere.

Yet another story concerns the Berlin Air Safety Centre, a popular place to visit for the Wives' Club. The centre is located in an eighteenth-century palace where, after the war, the original four-power government sat and where, during the war, Hitler installed the People's Court that was to order the executions, within its walls, of the growing queue of people trying to assassinate or otherwise work against him. Only half a dozen rooms were used by the centre, leaving the rest quiet and empty – the ideal place for ghosts who, though *super*natural, are, for some unaccountable reason, also shy.

The Wives' Club arranged one visit to it in the evening, when the place was quiet. They were shown the courtyard, where rows of bullet holes at chest level were said to recall how the enemies of Hitler were killed. The cells were alongside corridors from which the lavatory chains protruded: had they been inside the cells, the occupants might have used the chains to hang themselves. There was plenty for the wives' imaginations to work on. Sure enough, there was a power failure and all the lights went out, leaving the working areas only to be lit by emergency generators. Elsewhere was in darkness. Again, nothing actually *happened*, but the wives felt it politic to beat a diplomatic retreat rather earlier than they had intended.

Padres are sometimes brought into the battle on the supernatural plane. One padre recalled being approached by a white-faced young airman who told him he had seen a ghost. The young man had been sleeping in a Luftwaffe hut built for thirty men, but partitioned off by the RAF to produce rows of single units. The young airman said that, at one in the morning, an apparition of a German flyer in Second World War uniform had walked into his room through one partition and out of it through the other. He was wearing a steel helmet and carrying a gun. The padre asked the man occupying the room next to the airman's if he had heard or seen anything unusual last night. 'Terribly sorry, padre, I can't say. I'd had a skinful. I'll tell you how bloody drunk I was – I woke up at one o'clock and thought I saw a German soldier walk through the wall and out the other side.'

Yes, Berlin has a curious atmosphere, that much is indisputable; and the military unreality of its position may be a factor in that. Proceed well over two hundred miles westwards, taking the sealed train from West Berlin through East Germany and offering up your passport for minute examination by Russian soldiers, who also man all the passing station platforms; then drive past Gütersloh and through the smoking tall chimneys of the industrial Ruhr, and you will come to RAF Brüggen, and quite a different sort of atmosphere.

Here you are well away from the geographical front line, but nearer the world of political and military reality: unlike

RAF Gütersloh's Harriers, the RAF Brüggen Tornados can use nuclear weapons. This, among other things, makes Brüggen a prime target.

It is in the sharp teeth of this reality that life on RAF Brüggen is led. It is a much newer station than Gatow or Gütersloh, built in 1953 near the Dutch border and the Dutch town of Roermond. Its strike/attack roles used to be carried out by Jaguars, but during 1985 and 1986 these were replaced by one of the most modern and complex fighting machines ever devised. The Tornado is an all-computerized machine with two radar systems, ways of jamming enemy radar, a normal armament of thousand-pound bombs or anti-tank bombs, enough chaff (flakes of sheet metal) and flares to bewilder radar or heat-seeking enemy missiles and a laser marker to lock on to a target and never let it go.

The British taxpayer is getting a lot from RAF Brüggen, and paying a lot for it. The Air Commodore who was Station Commander at the time of my visit, a wiry high-flyer with energetic movements and expressionless eyes, told me about the financial implications of implementing the station's motto, To Seek and Strike.

Brüggen used to do reconnaissance as well, but these duties have been diverted to RAF Laarbruch, some forty miles to the north, allowing Brüggen to concentrate on its expensive strike/attack role. The aircraft and aircrew are valued at £748 million; the whole station is costed at £1,200 million. It is going to cost more. At the time of my visit, each of the four operating squadrons there, 9, 14, 17 and 31, had hardened aircraft shelters (HASs) for the dozen aircraft of each squadron, but technicians were still having to use old hangars of the 1950s. £50 million had been spent over the past five years on the shelters, and another £50 million or so was due to be spent over the next five years, if the politicians agreed.

Brüggen's role is vital. It is to watch an area stretching north from Kassel to the Baltic, together with the Belgian and the Netherlands air forces and elements of the Third

United States Air Force in the United Kingdom. 'We provide the front-line deterrent of NATO in the air,' said the Air Commodore.

Like Gütersloh, RAF Brüggen is a bigger station than most in the United Kingdom. It stretches seven kilometres east to west and three kilometres north to south, and has large explosives shelters to the south. There are 2,500 metres of asphalted concrete to the runway, and a similar length of taxiing way – more than is routinely needed. A hardened avionics centre was being completed while I was there, water and electricity mains were being improved, and the sewerage farm was being upgraded. Tornados involve more people than the Jaguars previously used. Each squadron has about 18 aircrew and 150 ground crew to tend its dozen aircraft. In addition, the station now houses 431 Maintenance Unit, which is responsible for maintaining the whole of the RAF in Germany, and which employs 650 people. The Air Commodore complained that it had to draw on engineering support housed in a 'tatty old wooden storage shed' at the moment; but he hoped that, too, would be rectified before long.

RAF Brüggen is a third bigger than the average station in the United Kingdom, and needs to be. At the time I was there, there were around 3,000 RAF personnel, plus some fifty from other forces, rather fewer United Kingdom civilians (mainly teachers) and four hundred or so local employed civilians (about two-thirds German, one-third Dutch) and about 4,000 dependants.

In a run-up to a war, the theory is that the dependants would be speedily evacuated, and nine hundred reinforcements from the United Kingdom, mostly tradesmen of the RAF, would be flown in, together with a reinforcement of the RAF Regiment which would guard the airfield with the help of German Territorials, and a contingent of experts on bomb damage. Bulldozers are already on stand-by at Brüggen for twice-a-year practices of how to repair bomb damage to the airstrip. This entails digging out loose rubble, putting in hard core and covering with pinned-down steel matting.

While trying to assure that its own runways stay open, Brüggen houses some nasty little ways of trying to close down the other side's. These include what is called the JP 223, a nicely impersonal name for a weapon which is dropped by a Tornado over an enemy runway. It spins off into a number of bombs, each with a parachute to slow it down so that the explosion, as it strikes the surface of the runway, takes its time to do the maximum damage.

'It poses the hell of the problem,' said the Station Commander, with some satisfaction. 'I would not like to try to clear our runway if the enemy have something similar. At the moment, they haven't.'

The main part of Brüggen's role is to counter air attacks, and to prevent an enemy moving up on the ground – by taking out bridges, road 'choke points' (points past which traffic *must* go) and rail facilities.

This role was radically enhanced with the arrival of the Tornado, about which the men of Brüggen rave. One said: 'The big step forward with the Tornado technology is that it gives us the ability to operate at night and in all weathers (unlike the Harriers of Gütersloh). I believe that NATO would wish us to use Tornado primarily at night. The enemy don't operate low-flying aircraft at night – generally speaking, they would operate larger bombers. We would deal with them.'

Training can pose problems at Brüggen, as at Gütersloh. The Tornados, on a day-to-day basis, practise low-level flying over the Continent and the United Kingdom. The Station Commander said that 'political restraints' on the Continent, and to some extent in the United Kingdom, inhibited night operations. They were not allowed to fly below 1,800 feet, which was not valuable training. The Tornados spent three-week periods at Goose Bay in Labrador, where they could fly at one hundred feet, and also had intense weapons training in Sardinia, plus exercises in America once every eighteen months.

I asked the Station Commander the £1,200 million question: how long would it take the Tornados to get into a war posture?

The Air Commodore chose his words with care, since it was not a subject on which one wanted to give a potential enemy excessively precise information. 'We don't anticipate we will be going to war with absolutely no notice. We could do it well inside twelve hours. We are talking about reaching three thousand personnel on the base, possibly in the middle of the night. We are talking about getting everyone into the nuclear–biological–chemical-secure area, securing the whole base, issuing weapons to everyone who comes on base, preparing all the war weapons, briefing aircrews and ground crews and putting the aircrew into cockpits as the aircraft become ready. When the final aeroplane has been prepared, then we declare ourselves ready to go to war.'

An interesting point about this explanation was that it included no mention of nuclear weapons, which the Tornado is known to have the capability of carrying. The fact of this capability is not denied, but the presence of nuclear weapons on any particular station is never confirmed or denied. This aids security and, perhaps, the peace of mind to those living on the base or near it.

'Recreational activities are vital to our job,' said the Station Commander. I could well believe it. There are provisions for glider flying and surfboard sailing off Kiel. There is a golf club, which serves the whole of RAF Germany. There are also two professional youth workers to help give sons and daughters of RAF personnel something to do with their spare time.

Humanizing the black nuclear question mark are a host of charitable and sociable activities. Between 1977 and 1986, RAF Brüggen raised about a third of a million Deutschmarks (about £100,000) to convert an old cinema in Brüggen town into a bowling alley. When I was there, the station was hoping to provide the finance for a community centre in nearby Elmpt. And RAF Brüggen personnel also put their cash into helping themselves. A few years ago,

£20,000 was raised to improve the open-air swimming pool, and £15,000 was donated to the RAF Benevolent Fund.

'It is a very social and united life, especially on the squadron itself,' I was told by the forty-year-old Wing Commander who was Officer Commanding 17 Squadron. 'This is the front line, so it is a very operational situation. You have people who have gone through a long and tortuous period of selection and training, and we have highly motivated, highly intelligent, educated people who love flying military aeroplanes. We are very much a family unit, all one hundred and eighty-six of us. You will fly together in peace and possibly die together in war.'

This is no protective sentimentality born of fear. It has a practical as well as a human value. 'We immediately get to know the strengths and weaknesses of all the aircrew on the squadron,' said the Wing Commander. 'A lot of people can *fly* an aircraft from A to B, but when it comes to *operating* it, they are slow or late.'

There was an inter-squadron competition, said the Wing Commander, and when you entered you had to edge your weaker aircrew aside. 'It is not quite an ego-trip,' added the Wing Commander hastily, 'but we have bombing ladders (tables). Over a period of time, we log every weapon we drop out of aeroplanes. If any crew malperforms, we investigate quite closely.'

As a pilot, the Wing Commander was obliged to live 'within the wire' of the station itself. This was so that his speed of reaction, if war were imminent, would be more assured. I asked him how soon he could be at the ready if the war caught him doing some peaceful gardening.

'I am on a short fuse here,' was the reply. 'If the hooter goes, I would expect to be into work in five or six minutes. If they were *all* gardening, every man jack ought to be on his location within one hour, if not on leave or duty elsewhere.'

Living in these conditions, could a pilot *ever* afford to get drunk? The Wing Commander said he could enjoy a drink or two on Friday nights, when he was relaxing. 'But the balancing of one's social activity to one's commitment is always at the forefront of our minds. I would obviously have a nervous breakdown if I didn't relax from time to time, but please don't regard me as a drunken alcoholic.'

The pilot's day can be long as well as demanding. Crews at Brüggen check in at six in the morning. Flying can go on until midnight. It is little wonder that a cool beer beckons on a Friday evening. This was so even at the time I visited Brüggen, when the seductive noises being made by the Soviet leader Mikhail Gorbachev, and the similarly placating noises being made by President Reagan, had led to the lowest security state for several years.

Like the Station Commander, the Commander of 17 Squadron of Tornados could not discuss the presence or possible methods of use of nuclear weapons. But he was able to respond when I asked him what were his *personal feelings* about life in a role which could include such weapons.

'Our major role will be with conventional weapons,' he said. 'There could be a stage in which we move into the nuclear role. It is an area that doesn't particularly bother us. Quite frankly, it doesn't bother me, not if Armageddon comes, because if we reach that stage, we know that deterrence has failed. The nuclear strike mission is going to be a complicated one, and one difficult to survive. However, we have put away the thought of Armageddon, because we have had peace for thirty years. I can assure you that, in training, we practise the procedures, though not with the weapons on board. It isn't something you can sit back and actually worry about. I don't lose sleep over it, neither do many of the fellows. They are combat pilots, they enjoy doing their combat. They find the technical challenge fascinating.'

Many of the pilots, perhaps, would be thinking more about their families than about themselves – worrying whether they were being taken safely out of the danger area and to the United Kingdom? The pilot acknowledged that, in war, there would be the 'odd guy who crumbled at the seams', but constant practice would steady most aircrew men.

'One is not selected for this because you are an unstable extrovert,' argued the pilot. 'We are probably stable extroverts.'

There was no room for gung-ho attitudes, he said. The special man for the job had an adrenalin that went with flying high-performance aeroplanes but was not a 'cowboy' character. The job was too serious for that.

I met one of the younger pilots on the station who, being newer to the job, might be expected to see the strains more clearly. He was a thirty-year-old Flight Lieutenant with the usual background for a pilot: early interest in model aeroplanes. Three things were on his mind, and none of them was nuclear weaponry. They were restrictions on flying time and conditions; married quarters; and paperwork.

The young pilot said he managed around two hundred and forty hours of flying a year. This was probably enough *only* because it was meant to *enhance* his flying performance, not establish it. Because of the amount of time that had to be spent on administration nowadays, there was not much time to do extra flying, though very new pilots were kept away from paperwork as much as possible so they could fly more.

I asked him if he thought there was too much paperwork. 'I certainly feel we are expected to take an awful lot into the working day. There are areas I could name as being over the top, as far as paperwork is concerned. The actual amount of time, pre- and post-flight, has become more substantial – perhaps one hour to every flight of about one hour forty minutes.'

Obviously his quarters were felt to be a lesser issue. He had a flat in Elmpt, about five or ten minutes away by car. He didn't find it satisfactory; but then he never regarded quarters as satisfactory anyway. They did not provide the standards one would expect outside, but you had to take the rough with the smooth. What he *really* objected to was the attitude of bureaucracy in relation to quarters. It was not that you were a customer who paid rent (though you did) and whose interests should therefore be looked after. It was, 'We'll take your money off you and see if we can fit in some of your wishes on the way.'

'And that isn't good enough,' he said crisply. 'But flying is why I am here, and it wipes out the various gripes.'

If flying hours are squeezed, so is the administrative and technical back-up. A tall, serious young Flight Lieutenant who was handling the back-up to 31 Squadron told me about the 'juggling' he had to do in getting aircraft serviced in time to meet promised dates.

Did he find the pressures *too* intense for efficiency, let alone comfort? 'The system as a whole is trying to squeeze a quart out of a pint pot. There is more pressure to generate flying hours with fewer and fewer resources.'

The result? 'You are effectively driving manpower harder. There is less leeway for niceties. There is less chance to allow guys to do the recreational things. Service-prepared sport *within* the working day should be one afternoon a week. I can't *remember* that! One cannot afford half a day. If there is an individual who wants to go on an expedition, even supported by the RAF, you have to look at the guy and say, "Is he trying to squeeze too much out of the system?" We have to say, "*Is* it character-building for him? How much time has he had off for other sports?"'

Getting spares in time to meet promised dates for the readiness of aircraft was sometimes a problem that added to working pressures, he said. The Tornado was a good aircraft to work on, because most of its parts were switchable on the bolt-in, bolt-out principle. But the system couldn't work unless you had the spares to bolt in. They often had to 'rob Peter to pay Paul' by removing parts from other aircraft which were in for servicing. In the long term, that might double or treble the work load.

This RAF aircraft nurse, I discovered, had come into the service at the age of twenty-six from Rolls-Royce, that most august of British engineering firms. The next question was obvious: did he regret the move?

'At Rolls-Royce,' he replied instantly, 'man-management was limited, and so was promotion. In the RAF, promotion and responsibility come quickly.' Like many RAF people, he always remembered one sober yardstick when grumbling about conditions: what conditions could be like in civilian life.

In civilian life, most people in control of a budget are aware of the Catch 22 situation when it comes to trying to save money on a good year: the only result is that you are trimmed back to that lesser budget the following year – which may well turn out to be a bad one.

But in the RAF, unlike civilian employment, there is little or no opportunity to pad one year's spending to safeguard next year's allocation; and the Catch 22 is more pressing. This I gathered from a Chief Technician in the aircraft servicing flight, a man exceedingly gnarled for his forty-three years.

And no wonder. 'We get pressures to produce aircraft in a shorter space of time, with too few men. The time allowed for servicing has been reduced in the last two years. The basic service of an aircraft used to be twenty-nine days, but it went down to twenty-seven. Because we are not taking as long to do the servicing, we are told that, as they see it, we don't need so many men. Therefore they chop the number of men. It is a vicious circle. It is an accountant's view of efficiency.'

Perhaps it is unfortunate that accountants themselves do not have to fight wars. After the Chief Technician, who had twenty-eight years' RAF experience, had said that equipment, as well as time, was in shorter supply, I asked him what he thought would happen in a real war. Would the servicing personnel be able to cope?

'To be honest, the RAF servicemen would adapt. If we didn't have the right special equipment, then we would adapt something else. You can do that in a wartime situation. Look at the Falklands. We did things there you could never get away with in peacetime. You do what is necessary to fly the aeroplane. The safety criteria don't go out completely, but they are there to a lesser extent than in peacetime.'

But if the serving men couldn't or wouldn't adapt, and laid down their tools? 'You take a gun and – bang!' said the Chief Technician, pointing a pistol-like forefinger at me. 'I am not saying we *would* do that, but it has happened, if you look at the world as a whole. It happened in Vietnam. But the most unlikely guy can turn into a real gem when you get him in field conditions in war.'

Let us hope so. I had formed a great respect for the way the RAF in Germany first candidly state the difficulties and tensions, and then set about overcoming them. It was an anticlimax, but also a relief, when I moved back to headquarters at Rheindahlen near Mönchengladbach, where the huge portrait in oils of the Queen on the dining-room wall of the officers' mess is presumably sufficient to make those who 'defend Dover at Helmstedt' forget the warped windows of the bedrooms, the rainwater seeping through vintage roofs and the timbers of the corridor windows rotting away in huge handfuls.

If it is not, the knowledge of the task at Europe's most crucial frontier must put into perspective the trials of trying to be a British RAF officer, let alone a gentleman, in the run-up to a possibly highly dangerous twenty-first century.

But the European frontier is not the only one that requires the muscle of the RAF.

3

NOT QUITE EAST OF SUEZ

'No, the British Sovereign Base Areas in Cyprus are *not* a state like Monaco. We have been shot at with mortars at RAF Akrotiri, a serviceman and a fifteen-year-old girl have been shot at during a two-mile car chase – that isn't Monaco. Cyprus is not the sunny, easy spot it used to be even three or four years ago. We have to keep our guard up all the time. My people are pretty tense.'

– Administrator of Sovereign Base Areas, Commander British Forces Cyprus and Air Commander Cyprus (an Air Vice Marshal).

'The general conception that a tour in Cyprus is a dream come true could not be further from the truth. It is hard work for long hours in very adverse weather conditions. For a privileged few, it is a nice tour. If you have got one of the better quarters – and most are substandard and need a lot of money spent on them – and if you have a car, you are okay. But for the majority of people, life is tough.'

– Officer Commanding Administration Wing, RAF Akrotiri (a Wing Commander).

Ask, if you dare, in the bamboo and chintz Episkopi headquarters mess – naturally as an agent provocateur – whether Royal Air Force Cyprus is in the Middle East partly to support the illusion that Britain still has clout east of Suez. The verbal missiles will arrive within seconds.

The game, you will be sternly told, is as serious or even *more* serious than it was in the days when Cyprus, island gateway to the Gulf states, flew the Union Jack, and the British were busy alternately imprisoning and releasing the wily, but ultimately unsuccessful, Archbishop Makarios II, advocate of union with Greece.

And what is that game now? Why do we need some 1,400 RAF men and women, and 2,300 men and women of the British Army, to stay in Cyprus? To protect the two Sovereign Bases which are British soil. And why do we need the Sovereign Bases? To enable the British forces to stay in Cyprus.

It is a circular argument and one which may be puzzling to taxpayers. They might not be quite so puzzled if they came to the realistic conclusion that the name of the game was ultimately Middle East oil and the containment of possible Russian expansionism via the Mediterranean; and that the men and women of the Royal Air Force, with their many radar domes and other communication devices, were arguably now its most potent and vital players.

The people of the RAF steer clear of secret details and policy matters as they go about their long duties on the three-hundred-square-mile island on which St Paul made conversions to Christianity in 45 AD, when Cyprus was still part of the Roman Empire. They concentrate on what they call the 'wall-to-wall sun' and (lately) the increased possibility of terrorist activities against them; ostentatiously they leave the wider issues to the politicians.

But they will point out that you do not need to be a strategic genius to see that if it ever came to a war between the West and the Warsaw Pact, Cyprus – though it has no NATO role – would give the West some options it would not otherwise have. The Lebanon is only forty miles away; Cyprus has already been used to evacuate Britons from there in emergency. Turkey, now a fellow member of NATO with Britain, is also only forty miles away. Israel is within sixty miles.

In an unthinkable but technically possible war, Cyprus could be used to make fairly deep air raids into the Soviet

Union. From it, the NATO southern flank could be defended, British naval ships could be protected, and the Russian passage from the Black Sea to the Mediterranean could be blocked.

Such sombre speculations do not detract from the fact that Cyprus is probably the most popular posting still remaining to the RAF man or woman, in a world which has shrunk severely for them since the contraction or disappearance of the British Empire, and in which English gentlemen no longer resolve every important matter to their satisfaction in every sunny part of the globe.

It is certainly not like that any more, though the bamboo furniture and reverential waiters in the verandah coffee lounge of the officers' mess of the joint services at Episkopi, seven miles from Akrotiri airstrip and a thousand light-miles away from a chilly and pushy Britain, might persuade the more credulous that it still *was*. Even the fact that the shanty-town shop centre at Episkopi joint services headquarters is called Dodge City rather than Tunbridge Wells does not completely erase the feeling of belonging to a sort of Raj: the Cypriot shopkeepers are so deferential, and even the local employees of the three Naafis 'Sir' you with a frequency not always to be found in Oxford Street or even Knightsbridge.

But, no, it is *not* like the Raj any more. There is a valley called Happy Valley, an enormous basin behind the south coast in which have been fashioned golf courses, polo pitches, sailing clubs, riding schools, an archery range and facilities for water skiing. But the men and women using them sometimes find their attempt at carefree laughter drowned in some rather nasty noises-off, the sort of noises-off that cannot in their turn be drowned, even in an excellent local wine at sixty pence a bottle.

There had been a few of these noises-off just before my visit. The 'alert' state was two notches up on the most relaxed state, known as Black.

An Army Corporal and a fifteen-year-old girl were driving from Episkopi in a Land Rover, towing a (fortunately empty) horsebox when they approached a group of men standing beside a parked car. The men opened up with automatic weapons when the Land Rover drew level and also threw a grenade, which missed. The Corporal accelerated as the other car gave chase, its occupants firing on them during a two-mile pursuit. Fortunately the soldier knew what he was doing and zigzagged back and forth across the road, simultaneously keeping the horsebox between the car and the terrorists and preventing the car from overtaking, when the gunmen would have been in a position to take better aim at the car and to block its progress. The couple escaped with minor injury. The horsebox was never the same again. Such a persistent attempt at murder must have removed in seconds, for the two in the Land Rover, the tonic effect of a Cyprus posting.

In another incident, a stretch of beach near Akrotiri airfield called Ladies' Mile was raked with automatic weapons fire. One person was slightly hurt. Other terrorists opened up with mortars, against both the beach and the airfield.

In all these cases, the terrorists were *not* Cypriots, but militants in a rather wider Middle East conflict, possibly connected with the Lebanon. That fact only enhances Cyprus's status as a place very much in the front line. It not only has its own tensions; other people's can spill over on to its soil. No sea can be quite so blue, no soil quite so red, no chalk cliffs quite so white, no vegetation quite so green, no sun quite so luminous. These are compensations. But even the apparent sunny goodwill of the average Cypriot cannot disguise the fact that an ex-imperialist Britain now operates in an ex-colonized world that can no longer be patronized as quaint.

There is a kingly job on the island that might delight Gilbert and Sullivan; one of the curious historical follow-ons from British imperial days which exist all over the world, though none perhaps more markedly than in the case of the Administrator of the (British) Sovereign Base Areas of Akrotiri and Dhekelia, Commander British Forces

Cyprus and Air Commander Cyprus. He was an Air Vice Marshal at the time of my visit, though the first two jobs alternate with a British Army General.

In some respects, as the Queen's representative, he is almost a king within his own fiefdom, although, in line with the British policy of respecting local sensitiveness, he played this down to me as rigidly as he did the sumptuousness of Air House, his official home behind rings of security-conscious trees at Episkopi headquarters. At first he stoutly refused to tell me how many bedrooms there were (later he admitted to five), and the nearest he got to the slightest suspicion of megalomania was when he admitted drily that, yes, Air House *was* rather more grand than the home of the leader of the land forces and that, yes, it *was* gratifying for a man of the air to be able to command soldiers.

The fact is that the dual role of leader of *all* British forces in Cyprus, not merely the 1,400 RAF personnel, gives considerable and real power to the man who holds the position. The fact that the man who holds it is also responsible for administering the British Sovereign Areas of Cyprus greatly extends that power to civilian matters within the two areas which Britain retained for herself as sovereign territory when she granted independence to Cyprus in August 1960 – after a particularly bloody campaign, by those vainly wanting union with Greece, to get the British out.

Together, the two Sovereign Base Areas occupy ninety-nine square miles. Except for the airfield itself at Akrotiri and the inner compound of headquarters at Episkopi, local people can live in, or pass freely through, the areas; but they are a kingdom within a kingdom. It is the Air Marshal, not the Cypriot politicians, who is responsible for the legal structure. He has under him a British judge who tries some four thousand cases a year. The British made the laws in Cyprus until it got its independence, and some of the laws in the Sovereign Base Areas are still British. One odd exception is that there is, as yet, no provision for the breathalyser in Sovereign Base law, because the surrounding Republic of

Cyprus (which is a member of the Commonwealth under the British Queen) makes no provision for the breathalyser.

There are many other anomalies which, in practice, make for partnership rather than friction between the British and the local Cypriots in the Sovereign Bases. Sixty per cent of the land is privately owned, some by Cypriots of the south and some by Turkish Cypriots. Only 20 per cent is Crown land and another 20 per cent Ministry of Defence land, covering areas like the airfield at Akrotiri.

But what the Cypriots can do with their sovereign base land is controlled. 'I allow agriculture and a bit of quarrying, which is now dying out,' said the Air Vice Marshal, a courteous, spry man with the face of a more reflective David Niven, the film actor. 'There are a few tourist cafés and restaurants. We don't allow any industrial development, which would snowball. I control the building permits. It is simpler not to have anything other than agriculture taking place where the military requirement is overriding.'

On the Air Vice Marshal's wall is a framed copy of his instructions or 'commission' in Cyprus Sovereign Areas, signed boldly in two-inch-tall lettering in the top-right-hand corner Elizabeth R.

'We do', she says 'by this Our Commission, under Our Sign Manual and Signet, appoint you the said' – here is entered the name of the officer – 'to be, with effect from the twenty-fourth Day of October, 1985, Our Administrator of Our Sovereign Base Areas of Akrotiri and Dhekelia, during Our Pleasure, with all the powers, rights, privileges and advantages to the said office belonging or appertaining.

'II. And we do hereby command all and singular Our Officers and loving subjects in Our Said Sovereign Base Areas and all others whom it may concern, to take due notice hereof, and to give their ready obedience accordingly.

'III. And we do hereby direct that this Our Commission shall determine upon signification to that effect being given by Us through one of Our Principal Secretaries of State.

'Given at Our Court of St James's this fifth day of August,

One thousand nine hundred and eight-five in the thirty-fourth year of Our Reign.'

Few RAF officers will ever have anything as grand to put into their scrapbooks. It is virtually a governorship over 4,500 Cypriot residents in the Sovereign Base Areas, together with 2,300 soldiers, 1,400 airmen, 400 British civilians on postings to Cyprus and some 3,700 Cypriot workers in the areas.

'If you compare this with the 1,800 people of the Falklands, you can see we are much bigger,' said the Air Vice Marshal. 'We are fifty times the size of Gibraltar.'

He thought that the job of Administrator of the Sovereign Base Areas went comfortably hand in hand with his duties at the military level. 'I find wearing the two hats a very natural thing. They sit well together. No one has ever suggested that we on the Administration are dictatorial or anything other than benign, but I suppose it *is* . . . not a dictatorship, but a sort of autocracy.'

Sometimes the occupier of stately Air House, with its royal coat of arms above the portico, finds himself in one capacity disagreeing with himself in the other. As an Administrator, he has to bear in mind that the British must be seen as a stabilizing influence. They have their own police force – three British officers and several hundred Greek or Turkish Cypriots, augmented since recent outbreaks of terrorism – who stop burglars and smugglers coming over the border between the south and the Turkish-occupied north. The Administrator has his own fiscal control system, and a British fiscal officer, who can board ships from a helicopter if they are in the territorial waters of the sovereign base areas.

The Administrator has a vested interest in operating with the least ruffling of the feathers of local people. This can be at odds with him in his *other* capacity, as commander of British forces in Cyprus. As a military commander, he would often like to seal off the Sovereign Base Areas in times of emergency to ensure the safety of his own men and women. But as Administrator, he knows this would curtail the social activities and the livelihoods of the local population. 'I have sometimes, politically, to overrule myself as commander,' acknowledged the holder of both posts.

He is aware that a political mistake could be expensive when there are local politicians expert at exploiting any differences between Cypriots and the British.

'We pursue a policy of neighbourliness, but there are some people who don't want to listen to you,' he said. 'If you look at the newspapers now, the Communist Party has nearly a third of the electors. If you take the Communists and the far-left Socialist Party, you are talking about more than 40 per cent of the population which votes for a platform which includes opposition to the presence of the British on the island.'

Relationships with the local inhabitants can generally be more touchy in Cyprus than, say, on the East–West German border, where the locals see the British as being there to defend them. The British are plainly in Cyprus to protect their own national interests and those of the West (though the bases have no NATO role, and Cyprus itself is non-aligned as between the West and the Warsaw Pact). If the Cypriots see the British as protectors, it is as protectors against the Turks, who invaded the north of the island in 1974 on the pretex of protecting Turkish Cypriots there. The Turkish Republic of Northern Cyprus was set up, not recognized by anyone except Turkey, and stopping short of the northern of the two British Sovereign Base Areas, Dhekelia.

At the time of the invasion, only about one thousand Cypriots lived in the Sovereign Base Areas. The number rapidly shot up to four and a half thousand as refugees from the north and south were allowed in. The bitterness of those who fled to the south is still great, and they still claim the north is theirs. They are encouraged by Greece, which is, however, further away than is Turkey, as well as being a lesser military power. So the final touchy reality is that the Queen's sovereign bases stand in a country divided by force, and are surrounded by Greek Cypriots supported by Greece and Turkish Cypriots supported by Turkey, both countries being members, with Britain, of NATO.

On reflection, Gilbert and Sullivan might have found the whole thing too kaleidoscopic and complicated to pull in the crowds; but it is the show that the 1,400 men and women of the RAF see every day of their working and leisure lives.

The situation of the northern part of Cyprus occupied by the Turkish is a quagmire for the unwary. The British, servicemen or civilians, risk intense anger if they, by word or deed, imply recognition of the Turkish regime. While I was in Cyprus there were three such cases, each causing inflamed comment among Cypriots.

Strong representations were made to the British High Commission in Nicosia about alleged 'confidential instructions' to British servicemen visiting the occupied north. They were said to be issued with 'travel permits' (local newspapers reporting these allegations always put those two words in quotation marks) issued by the Turkish regime, and were thus putting themselves in the position of recognizing that regime. The difficulties of terminology involved were made very apparent when one irate local publication referred to the 'breakaway Turkish Cypriot state' – thus itself implying that the northern strip of Cyprus *is* a separate state, something Cyprus has always denied.

The second cause célèbre, also involving a protest to the High Commission, concerned the practice of the British authorities on the Dhekelia British Sovereign Base, bordering the Turkish-held area, of operating an 'illegal' crossing point into the north. The allegation was that the British allowed base personnel and foreign tourists to use this crossing without the specific authorization of the Cyprus government, and without carrying out any checks to prevent 'smuggling or other illegal activities'. The British line is that it is legally impossible to stop British citizens crossing the border if they want to, for, as recognized by the legitimate government, the Republic of Cyprus is one and indivisible.

Case number three involved British forces only tangentially. It illustrated even more clearly than the other two complaints the way in which activities that most people would see as innocent can provoke anger in Cyprus. Pupils of Berengaria School at Limassol, ten miles from the Episkopi Sovereign Base but operated by the British authorities, visited Paphos, on the other side of the base area, to carry out a 'tourist enquiry'. It was reported to the police that young pupils were touring the town, asking tourists to answer a questionnaire of eighteen questions. One of them was: 'Would you have wished to visit Turkish Cyprus?'

A Cyprus government spokesman explained to the British High Commission that it was unacceptable for the occupied part of the Republic to be described as 'Turkish Cyprus' and that this constituted 'an act of provocation' to the people of Cyprus. The fact that the teacher and children had 'hidden' in a café once the police were told was held to be proof that they knew they were doing something wrong.

I need not comment on the rights and wrongs of any of these happenings – except to say that they illustrate with laser-like intensity the delicate line the people of the RAF, and their colleagues in the other forces and in civilian life, have to tread every day of their lives.

All get careful briefings on local politics before they go there, not so they can involve themselves in local politics but so they can *avoid* it. One pro-Turkish word in a café may be a word too many, and lead to fisticuffs. There are also battling factions within the south side of the island, some not friendly towards the British presence. Among the half-million people of the southern side of Cyprus (still held to be the *only* Cyrpus state by every nation in the world except Turkey itself), there are seven political parties. Three of them oppose the British presence consistently. The fourth opposes it intermittently, according to its coalition loyalties.

The Communists, Akel, are as predictable as their professionally produced and, it is said, Moscow-funded newspaper, *Haravghi* (Dawn). They are against the British presence, and promise the removal of the bases. Edek, the

Socialist Party held to be to the left of the British Labour Party of the late 1980s, also takes the same line. Both parties are adept at finding happenings in the news that can be used to berate the British. When both engines of a Lightning aircraft cut out, the RAF pilot stayed with his aircraft to avoid a village, ejected at the last moment and, though badly shaken, lived to attend a social function the same evening. But an olive grove was destroyed. *Haravghi* was quick to highlight the loss to the local economy and to ask how long it would be before someone got killed by a crashing British aircraft.

All RAF men and women in Cyprus have become practised in playing the economic card in reply to anti-British sentiments. Staying rigidly away from political issues, they will nevertheless manage to point out to their Cypriot acquaintances that the RAF and British Army are among the largest of the local employers, with 3,500 Cypriots on the payroll. And what do the British pay? Some 50 per cent more, in some cases, than the local pay rates for the same sort of job. The forces cannot afford discontent which would inevitably lead to a too-fast turnover in the workforce, equally inevitably leading to the employment of unsuitable and potentially disloyal people. When, in the main mess bar at Episkopi, I was presented with someone else's bar bill by mistake (a Mr Bower's), with the rather pointed request that I settle it as 'I' was leaving today, the apologies I later received were of the sort usually extended in Britain only, I imagine, to mega-millionaires and royalty.

The revenue argument is also used to defuse anti-British-forces feeling. British forces spend 60 million Cypriot pounds (about £80 million sterling) a year in Cyprus. There is also tourism, which often rests on connections with the British Army or the RAF in Cyprus. In 1986, nearly one million tourists visited Cyprus, some 400,000 of them British. It is *possible* these sorts of numbers would continue to come if the British forces moved out; it is by no means certain. Cyprus, it is reasonably pointed out, *is* in the Middle East, and British tourists feel more secure if they know that British forces are to hand – even to evacuate them if necessary, as British people have been evacuated from the Lebanon.

The British, as previous occupiers of Cyprus, have held on while other countries, once involved in the area through the United Nations, have backed off in discouragement at the apparently unsolvable problem of the north-south divide.

The border is divided into six sectors, with different nations under the UN taking responsibility for different sectors. But the Finns have withdrawn from their sector, closely followed by the Swedes. The Australians and the Canadians then spread their sectors to close up the gaps. But it has become obvious that the wider world is becoming restless about Cyprus's internal troubles, and that only the British, deft in diplomatic niceties, are wholly enthusiastic – while never, of course, forgetting their new aura of modesty.

'The Administration of the Sovereign Base Areas equates with a government,' said the Air Vice Marshal, 'though we are careful never to use a term that might suggest we have a residual colonial presence here.'

All that he would admit was that he had the power to make laws. It is also said that he, in the ultimate, has the legal power to impose the death penalty. This he would neither confirm nor deny, obviously finding the whole issue embarrassing at a time when capital punishment has been abolished in the United Kingdom except for treason.

It was thoroughly unlikely, I judged, that the Administrator would ever use his power of life and death, even if he does have it.

What Cyprus is partly about, in peacetime, is Intelligence and communications – about which little can be said.

RAF Akrotiri station itself is like no other RAF station in the world. At eighteen square miles, it is almost certainly the largest. It sits on the end of a peninsula of land, and the landing path passes over an expanse of muddy water called the Salt Lake, where pink flamingoes gather in the balmy waters. It has thirty miles of roads and sixty miles of storm drains – which are highly necessary in the rainy season. It has its own roll-on roll-off sea jetty, its own fuel depot, fed by pipes from the sea, and a big British Army Royal Electrical and Mechanical Engineer workshop.

But behind all these proud boasts is the sobering fact that RAF Akrotiri houses only a fraction of the permanent personnel it did in the early 1970s, before the Turks marched into the north. It has only one permanent squadron, which at the time of my visit was 84 Squadron, consisting of Wessex helicopters used for search and rescue work, often for Cypriot or other civilians in trouble. This squadron earns its keep in terms of local public relations, especially by taking into hospital women in premature childbirth. The Squadron Leader told me that this operation had once constituted 50 per cent of the workload, but in the past ten months there had only been one case, because doctors were now taking care to get women into hospital much earlier.

There are other differences between the search and rescue service the RAF operates in Cyprus and the one it runs in the United Kingdom. The helicopters, though also Wessex, are painted in camouflage colours rather than the more usual bright yellow; they may well be involved in hostile air action at any time. More importantly, 84 Squadron has to be self-sufficient in its ability to repair its helicopters without reference to the United Kingdom, though major parts can be flown in.

This search and rescue service, useful and gallant as it is, could be seen as an anticlimax compared with the picture in the early 1970s, when Akrotiri had Canberra and Vulcan bombers and Lightning aircraft to protect them, with a flying transport route into Iran. When the Greek–Turkish war made the route suspect and the fixed-wing aircraft went, the place was left, as the Officer Commanding, a Group Captain with world wide experience, put it, 'a real wasteland – it deteriorated very remarkably through the late 1970s, and was a really sleepy hollow in the perception of people in the United Kingdom.' Fortunately, he said, when the problems of the Lebanon had reached their height in 1983, Akrotiri could be used for famine relief; its other uses, and its strategic importance, slowly grew again.

As a transport airhead, Akrotiri deals with about four thousand passengers a month. There are four hundred transport movements. 'It is now the crossroads of the Air Force,' said the Officer Commanding, 'in the sort of way Gan and Malta once were. Now it is only Akrotiri. If you stand in the bar of the mess at Akrotiri long enough, you will meet virtually everyone in Strike Command on the flying side.'

Akrotiri is also a crossroads to the Americans. There are four US Army Black Hawk helicopters parked there, though operating out of Larnaca, to provide logistic support for their Embassy in Beirut. Akrotiri also provides support for the American U2 aircraft, which monitors the disengagement zones of the Arab–Israeli war.

It could not be expected that such activities, and such outside events as the American bombing of Tripoli in the mid-1980s, would necessarily lead to a relaxed and quiet mode of life for the eight hundred RAF personnel, two hundred British Army people and eight hundred Cypriot civilian employees who work on the station.

Since the Tripoli bombing, and the possibility of reprisals against UK interests by the Libyans or their supporters – which appeared to be realized in the mortar attack on the station in 1986 and the machine-gun attack on sunbathers on the nearby beach – the station has been subject to twenty-four-hour permanent observation-post duties. In this sense, it is perhaps fortunate that there is only a depleted workforce to protect, compared with the five thousand RAF people who worked at Akrotiri at its zenith.

In the midst of all these trials, Akrotiri has had to contend with that new-style British disease – lack of money.

The Officer Commanding had hinted at it – but the Wing Commander who commanded the Administration Wing spelled it out with brutal frankness, and with the verbal facility of a man who had put the painful argument many times before. 'One of the important things about Akrotiri is that there hasn't been a lot of money pumped into here for a long time. Because, at one time, the place seemed to have an uncertain future, the infrastructure has been neglected. The telephone system is falling to pieces, the sewerage is falling to pieces, *everything* is falling to pieces. We need a complete rebuilding of our married quarters, and we are talking about at least £30 million at today's prices.'

Not a happy state of affairs for Britain's continued presence east of Suez. At just the time Akrotiri urgently needed to have more money pumped into it to make up for the lost years, there had been a heavy *squeeze* on defence expenditure. When I was at the station, it had just put in for £2 million during the financial year for important improvement works, and had been granted just £200,000.

This money was likely to be spent on paint to make the place look superficially presentable, whereas, said the head of the Administration Wing, what was needed was major rebuilding work.

'This was a bad year; but we are not sanguine about the future, either,' he said. 'In the United Kingdom, corporals have a single room. Here they are three to a room. Married quarters are graded from one to four according to their quality. On most stations you have mostly grade one. At Akrotiri, it is mostly grade three. In my experience, this is very unusual. It comes down to poor square footage, lack of central heating, long distances from the Naafi, aircraft noise, heat and insects – these are some of the minus points taken into account when a quarter is graded. At every aspect where you have *minus* points, Akrotiri is able to score.'

The Wing Commander's conclusion was that morale was not *low*: it was *fragile*. People absorbed their energies in sporting or welfare activities. Without these the problems would be more serious, he warned.

Other officers bore out this argument. I was told that there had been 790 welfare cases – people who had admitted they had problems. Single people, especially, were apt to ask to go home early.

Single people do seem to have problems in Akrotiri. Their quarters are not comfortable. Since their taste of terrorism, they tend to stay in camp more. Since the pressures on the British to leave, local girls are virtually unapproachable; they always were jealously guarded by their fathers and brothers, and now only a very rash young RAF man would even try to get involved. There was, at the time of my visit to Akrotiri, no ban on personnel leaving base, but there were constant warnings about the dangers of going to the clubs of Limassol, ten miles up the road, and trying to chat up girls found there.

Yet some RAF men brave these dangers, goaded by the fact that, for the rank of Corporal or below, there are ten men at Akrotiri to every one woman. The young officers fare somewhat better: though the ratio of men to women in the RAF at Akrotiri is about the same, there are also women nursing officers and civilian women teachers who are also considered to have 'officer status'. This makes them acceptable people to take out for an evening.

Sometimes it seems to the sex-starved young RAF man at Akrotiri that the only available women are the wives of other RAF men, frustrated by the frequent working absences of their husbands and by the fact that Cypriots in Cyprus have priority, not unreasonably, in getting available civilian jobs, with the result that many British wives are unemployed. But any RAF man who gets involved with another man's wife in Cyprus would be sent home, whether he *wanted* to go home or not: the RAF can do without *that* sort of trouble, I was told in no uncertain terms.

Cynics might say that all this is a factor behind the RAF's passion for education in Cyprus. Nine hundred people go to various evening classes. The drop-out figure is less than 10 per cent, whereas among civilians, generally, it is nearer 30 per cent. Even the rambling club has sixty members.

For some, drink is the main release and a potentially expensive one, except in terms of money. The station's medical staff told me that some three hundred cases they treat in the course of a year – illness or injury through accident –

are related to excessive consumption of alcohol. With local brandy a few pence a glass, this may be understandable: in the United Kingdom, where drink is expensive, there would not be this number of cases, even in isolated stations in Scotland or the south-west.

The vast majority of RAF men and women at Akrotiri do not, of course, seek alcoholic escape: they merely suffer, though no longer quite in silence. The station has no swimming pool, which is regarded as a prime necessity in a hot and humid location where going to the beach means a long and irritating journey and a possible risk of being machine-gunned or mortared by terrorists.

In days gone by, there might have been pleas to the Officer Commanding. But just before my visit, some officers had made a home video film, taken from the passenger seat of a car during the long and hot journey to the sea, which set out the case for a new swimming pool. The video was to be presented to the mandarins of Whitehall direct in the hope of levering a bit more money from them. In the First World War, the film makers might well have been shot for something or other; today their superiors showed every sign of sympathy.

That sort of portent is perhaps disorientating for the traditionalists. There can be other disorientations at Akrotiri. Some of the officers, because of the comparatively low numbers at the station, wear two hats. A staff officer reporting to headquarters at Episkopi may also be a specialist officer, such as supply officer, reporting to a specialist Wing Commander at Akrotiri. One officer told me: 'Sometimes I write letters to myself, drive up the road and answer them.' Gilbert and Sullivan would certainly have loved this. I was reassured when told that, as the result of a review, the case for seventy-four additional posts at Akrotiri had been accepted. 'The RAF Personnel Centre at Gloucester will respond fully to this as soon as possible,' I was told. 'Men have been sent to fill some of the additional posts, but not others.'

The permanent personnel at RAF Akrotiri are only part of the station's working life. Squadrons from the United

Kingdom and other parts of the world visit it regularly for gunnery practice, known in the trade as Armament Practice Camp. It is a highly popular four or five weeks for the aircrews and their ground back-up. They can enjoy the sunshine in the spare moments from concentrating on gunnery single-mindedly. They can do neither in the United Kingdom, with its cold weather and confined air space.

Cyprus has a gunnery range twenty miles over the sea from the south coast, a strip some thirty miles wide and ninety miles long. Over this, Phantom fighters can fire away to their hearts' content without upsetting anybody, since Egypt is well over a hundred miles to the south. The firing range is not precisely to the south of Cyprus but slightly south-west, keeping it slightly further away from touchy Beirut.

I met 111 Squadron, whose Phantom fighters are normally based at Leuchars in Scotland, in the middle of their five-week visit.

The Wing Commander in command was an officer greatly concerned with morale. He had sought and received Ministry of Defence permission to have his own Phantom painted partly black, with a yellow flash on the tail bearing the squadron badge, his name painted on the side of the cockpit and that of the navigator on the side of his. This was *not*, he explained to me, meant to be ostentatious but to help rally the other fighters to the cause.

He was a tall bald man with a handshake that could cause fractures, and a patent enthusiasm for training conditions in Cyprus.

'We have various exercises throughout the year, like missile firing off the west coast of Wales or the Outer Hebrides,' he told me. 'But we can never do it as a complete squadron. Here we can be a complete squadron and a squadron blinkered on gunnery.'

Thus some two hundred and twenty airmen and Leuchars station personnel, with fifty officers, were at Cyprus – the largest fighter squadron in the RAF completely uprooted and set down in an exotic climate to shoot holes in a twenty-four-by-six-foot flag towed a safe nine hundred feet behind an old Canberra bomber.

The speed at which the pilots have to work to hit this 'enemy' flag is tremendous. The young men taking part had instructions to open fire at four hundred yards from the 'enemy', while overtaking it at over two hundred miles an hour. This gave them half a second in which to fire in order to score a hit somewhere on the flag. In that half-second the gun fired over fifty rounds, any one of which might be enough to shoot down an enemy aircraft, especially as the aim is to hit the back of the neck of the enemy pilot.

Each pilot attacks the flag in succession. It was not until I inspected the Gatling gun they used that I realized how it could be established with certainty which one of the pilots had managed to put the holes in the flag. These particular Gatling guns are suspended like pods under the fuselage, with six barrels capable of firing one hundred rounds a second. The rounds of the one I saw were painted green with a plastic paint that never quite dries. Those on other Phantoms were painted different colours. Each hole in the flag, at the end of the firing, was surrounded by a tell-tale ring of paint in a colour betraying who had fired the shots.

While the pilots fly, there are opportunities for the control tower staff, some of them women, to cope with the problems of guiding and talking down the Phantoms. One pilot found that his training parachute did not open properly on landing, and probably would not have pulled him up before he reached the end of the runway. He jettisoned it, rolled, took off again and circled until he had lost enough fuel to risk a landing without the trailing parachute. He got a friendly wave from the control tower as he finally swept past it.

The wave came from a woman pilot officer who was talking down the aircraft, a set of headphones on her head, a cigarette between her lips.

This was obviously high-tension work. 'You can give pilots advice, but they don't always have to agree,' she said between puffs. 'I didn't smoke until I came out here – but you'd better not write that down or my husband will kill me. I'll tell you this – in this job, you can't afford to have a hangover. I tried it once. Never again.'

In fact, work in Cyprus can be less exacting for the control tower personnel than it is back at their home base. There are fewer aircraft movements to be dealt with in a day, and the women have time to sympathize with the Canberra crews, who are not attached to individual squadrons but spend most of their working lives on the Cyprus gunnery range, flying more or less straight up and down the length of it. There were friendly waves for the Canberra crews, too. One did not venture to ask what went on after working hours.

But work is the essence of these sort of visits. Apart from the training, both the resident Canberras and the visiting Phantoms can be diverted to look at any strange shipping in the area. Normally such ships are ordinary fishing vessels. But the reported belief that a high proportion of terrorist activity in the Middle East has a Cyprus connection makes aircrews at Akrotiri, resident or guests, constantly alert.

Alertness in gunnery practice is the most important of all. At the end of a series of sorties, I saw a film taken from a camera in the nose of one of the aircraft. Such films are always played back for the benefit of the pilot concerned, who, on the first two passes across the enemy in the sky, 'shoots' only with the camera itself. Thereafter the firing is for real, the shots also recorded on the film.

The RAF is reticent about discussing the exact pass rate for shots hitting the target, but a pass mark of about one shot in five would not be unrealistic. Many pilots get far higher percentages of successful shots: the most successful on the score card I saw was 53 per cent.

The hair-trigger work of the visitors is backed up by no less demanding but less glamorous work of the station's resident personnel – especially the engineers and the firemen. The engineers, like the personnel in the control, can argue with pilots but not, in the end, overrule them. This situation can seem markedly pointed when the Officer Commanding the Engineering Wing at Akrotiri is away and a more junior officer has to deputize for him. One twenty-nine-year-old Flight Lieutenant pitchforked into this delicate position told me that he was responsible for engineering standards on the station, and that if an aircraft's ground crew wanted to do welding work on an aircraft in a hangar to keep it flying, he might have to veto it if it could (in his opinion) lead to a fire that could burn out the whole hangar.

'You could have millions going up in smoke thanks to me,' he said. 'I might have to say, "You will do the work elsewhere on the station or fly away." But most of the time it is standard procedure, and quite straightforward.'

Most pilots of today's highly technological and expensive jets are quite willing to play safe in peacetime conditions; the wizard-prang mentality is out of fashion. One pilot encountered a technical problem when two hours out of Akrotiri, turned round and flew back for a ten-minute repair job, since all other airfields with the right technical back-up were closed for the night. Even Akrotiri itself has a limited range of spares. In wartime, no doubt, the spares in store would be increased at Akrotiri, while other airfields would condescend to stay open rather longer.

And the firemen? Apart from handling about two thousand call-outs a year on the station, including cases where aircraft arrester parachutes fail, they go out into the surrounding community to help extinguish about one large grass or scrubland fire a week. Some of the fire-fighting vehicles they use are thirty years old, but I was told by one of the fire-fighting personnel: 'We love them. We have to go to so many bush fires, but if you scratch one of these vehicles, you just slap a bit of paint on and it doesn't notice, or matter.'

A good example, surely, of the British stiff upper lip – which can still exist in tropical climates where once the British ruled supreme. It is only on what, to foreign observers, would seem minor matters that the British calm is allowed to evaporate slightly.

When I was in Cyprus, an R A F competitive drama festival was in progress (the R A F Theatrical Association has over forty affiliated clubs). The titles of the three contending plays quite elude my memory, but I shall always remember the familiar virulence with which the losers speculated on the parentage and mental capacity of the adjudicator. It could have been a drama festival in Woking or Wolverhampton.

In Cyprus, there is a foreign field which will be for ever England. Or, if not for ever, at least as long as even the militant Cypriots regard the R A F as preferable to the alternative evils.

In other overseas locations, there may be less diplomatic hassle; but there are still challenging problems, not all of them military.

4

Furthest Outposts

> 'Morale at Mount Pleasant is extremely high, but I have done two tours in the Falklands, and I have known men on their fourth and fifth tours, and it is going to happen again and again. Four months away from the family, 8,000 miles from home, in the difficult conditions, with the lack of recreational facilities – it isn't one of the most popular things going on in the service. Could we go on for ever? Yes, we could.'
>
> – Commanding Officer, Mount Pleasant RAF Station, Falkland Islands (a Group Captain).

> 'Belize, though independent, was once part of the British Empire. That adds to its attraction. And we are welcome there. The Americans want us because we are not Cubans, the Cubans want us because we are not Americans, the Guatemalans want us there because we are not Americans, the Belizeans want us there because we are not Guatemalan. I sometimes think the only people who want us out are the British Ministry of Defence.'
>
> – Senior Officer, RAF, British Forces Belize, Central America.

The irritating but survivable riddle of the Falkland Islands for the men of the Royal Air Force is that when conditions got better for them life became arguably rather worse.

The primitive huts around the old airfield at Port Stanley, the only town, and the virtually round-the-clock duties amidst the icy gales of the South Atlantic, produced strain.

The comfortable isolation of the £400-million newly built military complex around the new airport, thirty miles from the 'civilization' of Port Stanley, sometimes produces boredom.

Some men find the boredom marginally worse – though rarely, through either strain or boredom, do any of the thousand or so RAF personnel (exact numbers are never stated) collapse and have to be sent home. It *has* happened that an RAF man starts to 'behave strangely', is reported by his mates and is sent back from what one man called 'the ponging penguins' to the more ordinary amenities of the United Kingdom; but for most men four unaccompanied months watching for an Argentinian attack which never comes is *just* compensated for by the exotic wildlife and the cash savings that can be accumulated, since there is very little to spend money on.

In the immediate aftermath of the liberation of the Falklands, life was constant toil amidst dust and grit, snow and gales. Men who slept six to a ten-by-six-foot hut considered themselves lucky to have any sort of solid roof over their heads. Tents were the norm.

The old airfield at Port Stanley was the obvious focus of the military presence, but it had only four thousand feet of runway. This was sufficient for the Islander aircraft of the Falkland Islands Government Aviation Service, but not for the VC10s which at that time – 1982 – were the RAF's main jet transporters, let alone the later wide-bodied TriStars. The airfield had to be turned into a military base almost within hours; and within days it became a military base crammed with aircraft, Rapier surface-to-air missiles, other military hardware and British Army units – including infantry units, who acted as building contractors as well as potential defenders of this British territory a third of the globe away from the United Kingdom.

One RAF technician remembered sleeping in a tent of canvas with metal supports. It hardly mattered, he said, because he spent most of his time outside, where the work then had to be done. 'The conditions were appalling,' confirmed an officer, on his second tour. 'There was a lot of mud and smoke and dust. Most people had at least a twelve-hour working day, some much, much longer. Morale was fantastic – there was a job to be done and the boys got stuck in.'

Conditions began to improve, slowly, when a ship was moored in the nearby harbour of Port Stanley, soon to be replaced as the main accommodation centre by a Coastel, a multi-decked floating house-ship designed to take two hundred and fifty oil-rig men. Standing off Port Stanley harbour, it held one thousand servicemen. 'It was crammed, it was hot, it was noisy – it was *luxurious* to what they had been used to,' said a senior officer who was there at the time of the transformation. Even here protocol was observed: officers had one dining room, airmen another.

In the days of the original dormitory ship, about the size of the Royal Yacht *Britannia*, it often took men nearly an hour to get into work. They had to take a small ferry-boat to Port Stanley jetty. This would have been dangerous as well as time-wasting had there been another Argentinian attack – and was unpredictable even without one. The water was often so rough that men were sick; sometimes the sea was so bad the ferry-boat could not land at all.

Even if the ferry-boat did land, there was another hurdle. The 'road' – hardly more than a track – from Port Stanley town to the airport was full of potholes and covered in mud. A journey of some ten miles took forty minutes, after which the men had to reach their individual places of work on a very scattered station.

The Coastel was tied alongside the shore nearer the airport, but the route was just as bad, and could only be handled by Land Rovers, often using four-wheel drive, or four-ton trucks. Men hiked into work and home again in the evenings, and kept careful watch on one another: if a man fell out of sight in a ditch, he would have died of exposure. Neither were conditions aboard the Coastel *objectively* luxurious: airmen slept six to a small room with all their clothing and kit. Not, remembered one technician, that junior ranks took much in the way of clothing with them, only combat gear and perhaps a pair of trousers and a sweater for 'rest' periods, if there were any. At the end of the corridors was the single ablutions room. Men queued up to get into it.

Yet officers remember the place as a hive of activity with morale very high as men sweated to put roofs over their own heads *and* to get Port Stanley airport ready as a full military base for the defence of the Islands. One said: 'Everyone was conscious of the threat, which at that time still seemed to be a reasonable threat, bearing in mind that there had been no formal cessation of hostilities – still hasn't been, for that matter.'

Everyone knew what the main priority was: the defence of the airhead, which was the airport itself. There were Phantom fighters and Harriers, two or more of them always ready on Quick Reaction Alert to take off at a moment's notice, just as back home they would do in defence of United Kingdom air space. They were backed by Hercules aircraft, acting as freighters and tankers. Helicopters, including Army Gazelles, Sea Kings and Chinooks, supported the forward radar and communications sites. On one or two sites, like Kelly's Garden and Navy Point – no more in fact than collections of portacabins – there were self-sufficient helicopter flights.

While flying was going on, the Royal Engineers took on the job of lengthening the airstrip itself to the six thousand feet needed by the Phantom fighters. This they did by putting metal mesh on top of the old surface. A quarry was started. Blasting produced rubble for use on metalling the mountain tracks around the airport. At that time, there were some four thousand servicemen on the Islands.

For months, the only way men or supplies could get to the Falklands was in a Hercules, a versatile and steady propeller aircraft which is both noisy and uncomfortable, requiring the use of ear plugs. Other aircraft could deposit men and supplies at Ascension Island, but everything had then to be put on this Hercules air-bridge, which entailed a fourteen-hour flight. Often Port Stanley was covered in fog; and the Hercules then had to return to Ascension. The air-bridge was a vitally important but tenuous link: ships from Portsmouth, bearing other supplies, took over a month to reach the Falklands.

The Phantom pilots did combat air patrols to police the Falkland Islands Protection Zone which Argentinian warships and military aircraft were forbidden to enter. Their work was relatively comfortable. Those airmen who kept the aircraft serviced had to work on them outside in the driving rain, braving a nasty circumstance called the chill factor. There was no more snow than would fall in the United Kingdom. But temperatures of zero or below, plus winds of thirty miles an hour, were the norm: they made it virtually unbearable outdoors after only a relatively short period of exposure. Men working on aircraft found that tools slipped in their hands. Quite a few were injured because of this. There were cases of hypothermia and frostbite.

A collection of portacabins had been set up at the airfield, and called a medical centre. It had no beds – to be fully hospitalized, you had to be taken to Port Stanley town, where a British military hospital had been set up, or to Port Stanley's civilian hospital, which was to burn down in 1984, through an electrical wiring fault, with the loss of some elderly patients. Because the ground was so rough, and because men had to do a lot of walking over it in the absence of better transport, broken or sprained ankles were common. I was drily told that there were no 'health and safety at work regulations' in operation. If there was a job, the job got done – if necessary, in frankly unsafe conditions.

Quite a lot of injuries in those days were unavoidable, some were not. The Argentinians had not mapped their minefields, designed to kill or maim as many of the British forces as possible. They had also left behind vast quantities of mortars, shells, bullets and other ordnance. In one week, an Army major had his foot blown off when trying to defuse a bomb, an airman had half his hand blown off when meddling with something he had found, and someone else had half a leg blown off when stepping near a mine.

An officer told me that in those early days after the campaign, his only relaxation had been occasional fishing. For this, he had to drive his Land Rover, on four-wheel drive, some two or three miles, and then walk for forty-five minutes through what had been the main battle area: 'It was like a rubbish tip. There would be an Argentinian tube of toothpaste, and various discarded clothes, plus a large amount of mortars and small-arms ammunition.' He picked up the live small-arms ammunition, and always came back with a bagful. He reported the position of the mortars, without touching them. For the year of his tour, this was the nearest he came to relaxation.

'But the biggest single hazard was overwork,' he said. 'Everyone worked long hours in poor conditions, and their spirit got them through. A bit of sense of danger which, I suppose, military people like.'

At first he refused to be drawn about whether there were people who had cracked under the strain. Finally he said: 'There were, but I wouldn't like that to go out as a blanket statement. There were a few, but they were a very, very tiny proportion, the people who found it too much.'

One Squadron Commander of a Phantom squadron remembered having to send one of his men back to the United Kingdom. The man had been – he would not specify in too much detail – on a very important duty, very important to the safety and serviceability of the Phantoms. He started to say 'funny things', which had led his colleagues to think he could intend to injure himself, or other people. Since he was in a position where he *could* have done such damage, his colleagues reported his rambling threats, and his Commanding Officer took the view that he should be removed at once from a situation in which he could do himself or others harm.

The important point here was that this Commanding

Officer could still remember the incident well, after a number of years, precisely because it was so rare.

The more usual pattern of behaviour at that time was for officers and airmen to air their reservations about the available accommodation, but then go on to rave about the geographical glories of the area.

'I was one of the few people on my squadron to have a single room to myself,' remembered one officer. 'Most lived and worked in two- or four-men rooms. Accommodation was minuscule none the less – I lived in a room seven by ten feet in the Coastel, and thought myself lucky. The only time we could get off that Coastel was when we were on Quick Reaction Alert duty, and that meant sleeping in sleeping bags.'

But, said the officer, you had to open up your imagination to the landscape, which was remarkable. 'The Falklands has the clearest atmosphere I have ever seen in my life. When the sun comes out, it burns people immediately, it is so clear. I can recall one sunset that was far and away the most beautiful I have ever seen. It was as if the sky were on fire. Very orangey reds, filling not just the horizon, but at least thirty or forty degrees above the horizon – covering it with this incredibly vibrant orange, the effect of the sun and the clouds. The quality of the atmosphere, I am sure, had a lot to do with the brightness and vibrancy of the colours.'

Even this enthusiastic officer conceded that, in the old days, men would not have been able to go on in these conditions for ever. That was why everyone was on four-month rotation.

At the time I spoke to this officer, the argument was still raging about how long the *ideal* tour of the Falklands was. For an experimental time it had been lessened to five or six weeks instead of four months, but the snag was that men would have to do *more* tours of the Falklands if the same presence was to be maintained. Some were doing three short periods of duty in a two-year term, and finding it enabled them to enjoy neither their families in Britain nor the unique atmosphere of the Falklands and their people.

'Do you prefer a big smack on the head or two or three little slaps on the wrist?' said one officer. 'That is the question some men are still asking themselves.'

By the mid-1980s, the conditions of life had greatly changed for RAF personnel in the Falklands, though the duties remained onerous and the sense of real threat was hardly reduced, since the Argentinians had still not officially declared an end to the war they had launched.

Men who had served in the Falklands in the early days could hardly believe the difference, however. The whole military complex had been shifted from Port Stanley airport to two new airstrips in a completely new military building complex at Mount Pleasant, thirty miles away. On what was virtually a peat bog, a British consortium of contractors had built an airport of international standard, and an RAF station with its own hangars, a few small hardened hangars and workshops. A video called *80 Weeks to Touchdown* recorded how the bog was changed in that time to a virtually completed project, requiring its own supply harbour, three miles from where the airfield was going to be, and its own road from the airfield to Port Stanley.

The main runway is now nine thousand feet in length, ample on which to land a huge TriStar aircraft in which the RAF now flies its personnel and relatives from Britain via Ascension – some of them grumbling about the military red tape of the flight, on which the passengers at Ascension are told, for instance, that anyone possessing alcohol will be taken off the flight, 'at the next stop' – which happens to be the Falklands, precisely where they wanted to go in any case. 'The RAF must have irritated a good many civilian taxpayers in this way,' said one civilian who had made the journey several times. He was in a minority: most people who made the journey by Hercules in the early days regard the present conditions as almost sybaritic, deflecting attention from what is still seen as a real military threat.

The cross runway on which smaller aircraft can take off, if necessary, is five thousand feet long, and there is a network of taxiing ways.

But one of the most startling sights, to a visiting civilian, is the completely covered domestic sleeping quarters, including its own shops and gymnasium, part of it two-storey, other parts single-storey. I said I was tempted to call it Mini-Soho, and was sternly told I must do nothing of the kind, as mothers and wives of RAF men in the Falklands must *not* be led to think that there are the same amenities for their men as you would find in Soho.

I am on safer ground if I say merely that it is impossible to say what it resembles, because it doesn't resemble anything else. It is strictly a one-off. Go through the main entrance and you face a sort of huge corridor down the middle of the complex. This is known colloquially as The Street, and is a half-mile long, with other 'streets' off it, leading to the bedrooms, the restaurants, the gymnasium, the bars and the church. When off duty, everyone can live indefinitely without setting foot into the open air.

This is just as well, as the air, as well as being fresh, can be cold and damp; and the roads, though improved, can take a heavy toll on RAF vehicles and RAF men's feet.

The Commanding Officer of RAF Mount Pleasant, a Group Captain, now has his office in the station headquarters, a sectionalized building in the middle of the complex of air base buildings to the north and south of the whole site.

'The roads around the airfield itself are adequate, but the roads outside it are not like the M1,' he told me. 'The toll on vehicles is still very great. There is still a small contingent of men at Port Stanley, and the road to that is *called* a road, laughingly, but sometimes it isn't much better than a dirt track. I dread going to dinner with the Governor, not because he gives an awful dinner, but because the road to his residence – though the islanders think of it as the M1 – is such that if you burst a tyre, you would probably kill yourself.'

One of his signalmen had in fact been killed when his Land Rover overturned on ice on one of the airfield's roads.

But, said the Group Captain, the problems at Mount Pleasant were not usually lethal. 'Whilst the domestic accommodation is much better at Mount Pleasant, some of the problems have been taken with us. There is nowhere for the men to go, very little in the way of recreation. If a man is interested in wildlife, the Falklands are second to none – except that you can't get there, anyway. Because of the torrential rain, you don't get very far – a five-mile walk is quite a long way, believe me. Whereas at Port Stanley you had everybody active, with very little time to spare, but able to go occasionally into Port Stanley to look in the pub, The Globe, and have a beer, in Mount Pleasant there is nowhere for them to go. And there is a little *more* recreational time, not because we are not working hard, but because the workshops are now so much better that there is not the need for the frantic activity there was at Stanley.'

Money and thought have been put into solutions. On Gull Island Pond, really a lake between the airfield and the harbour, there are now facilities for canoeing, windsurfing and sailing – but often the winds are too brisk for even the hardiest spirits. Hockey and rugger pitches have been carved out of uneven ground – and are still pretty uneven. The day they were officially opened happened to be a fine one, and things went well; a few days later the wind was so harsh that it was impossible to stand up on the pitches, let alone direct a ball. A number of rest and recreation centres have been set up – a rather grand name for a number of small cottages leased from the local settlement manager and equipped with a number of beds and a cooker. Most are near places where there is interesting wildlife. There is a system by which, if there are spare seats on helicopters, men will be taken to these centres for a few days' rest. They see penguins at six-feet range – if they can stand the smell, which is a cross between fish and ammonia. It appeals to the older men rather more than to teenagers, a few of whom make rather too generous use of the adequate number of junior ranks' messes and bars. Surveys, however, uncovered the reassuring fact that alcohol-related offences were fewer than on stations of comparable size in the United Kingdom.

Perhaps everyone, from the most junior ranks upwards, has an awareness that they have a real job to do in defending the airhead, whereas in Britain there is less of a perceived threat. There are fewer men at Mount Pleasant than there were at Port Stanley, though just how many fewer is not disclosed. The plan would be to reinforce with more men in the event of another emergency; but this plan, too, is not disclosed. Neither is anyone keen to discuss the RAF's exact tasks in looking after the out-stations and radar stations on the islands, except to say that British Army Infantry are there in force to guard against possible attack.

Unlike most airfields occupied by the RAF, which have grown up piecemeal, often against the rules of logic, Mount Pleasant is a more rationalized community, with the RAF squadrons spread around one side of the airfield, and the technical, supply and catering empires on the other side, ending in the huge domestic complex, set well away from everything else. Headquarters sits between the squadrons and the other facilities, a two-storey building of pre-constructed panels, but looking permanent enough. It is a joint forces operation, with a two-star British Army General, the only officer senior to the RAF Commanding Officer of the station, having control of ships as well as the troops.

The RAF Commanding Officer has control of air defence, the helicopters, the Hercules and Phantom aircraft and the RAF Regiment. His task is broadly to 'listen' to the forward-placed radar stations and send up a quick-reaction Phantom – or better still, two – to identify and investigate any aircraft or ship that looks as if it might be about to enter the protection zone around the islands.

The Commanding Officer said this would happen two or three times a week. He was cautious about explaining the exact nature of the events, except to say that what they were more worried about was the Argentinian aircraft coming near the Falkland Islands Protection Zone. In general, the Argentinians seemed to respect the zone itself, but often came fairly close to it.

I asked the Commanding Officer whether conditions in

this respect seemed to be improving or not. 'I don't think so,' he said carefully. 'When you look at the nature of the Argentinian man, who is a volatile chap, the state of politics and the economy in Argentina, the dirty wars against human rights of the past, with no one above the rank of Colonel able to be tried for such things, the extremely great power of the military and the very difficult and complex relationships between the civilian government and the military, I don't think you could expect to say that things are easing off and getting better.'

Were there many aircraft or ships which innocently went near, or into, the protection zone? None of it was really innocent, said the Commanding Officer, that was the trouble. There was an air route from Chile to Rio de Janeiro, from which it was possible that aircraft might stray east towards the zone, but that had not caused the most trouble.

We were now obviously getting into areas of understandable caution about the workings of the Falklands presence. I asked the Commanding Officer if there were ever signs of Cuban or Russian interest in the zone. 'One cannot talk about that,' was the only answer.

But he said that everyone realized that, though Phantoms were on Quick Reaction Alert at Mount Pleasant in physically the same way as they would be in the United Kingdom for the defence of the home air space, with pilots and navigators of at least two aircraft always in their flying gear and within a few feet of their aircraft, at Mount Pleasant there was a higher alert state than would apply normally in the United Kingdom.

'Whereas at a place like RAF Wattisham you are really checking on the air space, and policing is almost too strong a word, if you scramble a Quick Reaction Alert at RAF Mount Pleasant, you are in a potential threat situation,' I was told. 'That is felt all the way through, from the chap serving meals right up to the pilots in their flying gear waiting to react to a real threat.'

As in Cyprus, the top RAF man in the Falklands has a

partially diplomatic role, as well as his military one. He entertains visiting officers and politicians, local Falkland Islands officials, visiting industrialists and representatives of the other services.

Or perhaps one should rather say that his *wife* does. It is in the area of entertaining that women, conspicuous by their absence in other aspects of Falklands military life, come into their own. There are only nine married quarters in the Falklands, for the wives of senior executives – Wing Commanders or a corresponding rank – *only*. The tours of such senior executives can last a year, whereas men without their wives serve only four months at a time if junior officers or men, or six months if top executive officers. Though there are nine married quarters, it is rare for them to be all occupied at any one time, especially as some senior officers, who would be entitled to bring their wives with them, do not in fact do so. Either they are reluctant to leave the United Kingdom, or there are children doing their education *not* at a boarding school, or the officer decides himself that the Falklands are no place for a woman and that he will do his job better if he does not have to worry on her account.

Seven wives are quite a usual number. This little group spends much of its time entertaining and being entertained, a situation which officers frankly admit can be psychologically incestuous though, on the whole, good for morale.

If the wife of the Commanding Officer of the station prepares for a dinner party in the Falklands, it is rather different from planning it on bases in Suffolk or Lincolnshire. There are no convenient delicatessens from which tasty morsels can be purchased. The wife has to trek to the rations store, run by the British Army, though under the control of the RAF Commanding Officer as part of his Supply Wing. The store is almost the size of a hangar, and is full of food freezers the size of an ordinary domestic room. Serving staff will normally hack off a joint of beef of manageable size. Most other commodities have to be bought in large quantities – most of them in plain, uninspiring tins about two feet tall. This includes jam and marmalade. Tins of these tend to last for months.

Choice is inevitably restricted, since everything has to be shipped or flown in, much of the meat from Uruguay, in the absence of supplies from Argentina.

There is an even more restricted range of fresh vegetables. I asked one officer what vegetables could normally be obtained for a dinner party. Potatoes were available in bulk, he said. Carrots? Only in tins. Broccoli? None. Courgettes? None. Peas? Only in tins. Green beans? No. Baked beans? In large tins only.

Butter and margarine come in manageably sized packages, and bread is baked on the station, but fancy sweets that require cream are sometimes beyond the limits of the system.

And the Commanding Officer's wife, or other senior officers' wives, does not give dinner parties twice a month, as she might do in Britain. She gives them probably twice a week. It is just as well that most faces at RAF Mount Pleasant change every four months, so that, in a one-year tour, the Commanding Officer and his wife will see three sets of junior officers and perhaps two sets of senior officers and their wives. Otherwise some men and their wives might begin to brood that even hostile Argentinian faces on the horizon would at least provide *variety*, a commodity in short supply in this little remnant of the British Empire which suddenly became an international cause célèbre and a training ground, at very least, for defence strategy and tactics.

For the men of the RAF, the *reality* of the Falklands operation is clear-cut. British people are involved, some two thousand civilians – farmers, engineers, builders, hotel keepers and their wives, even a solitary journalist, who runs the *Penguin News* newspaper. Some were born there, others elected to go there.

The RAF has endeared itself to civilians in the Falklands in a number of ways, official and unofficial. Personnel, since the move to Mount Pleasant, tend to be even more inconspicuous than those of the other services, both in personal manner and in terms of the fact that they probably have less reason to come off station than Army and Royal Naval men. Personal manner? One Army officer confided to a local resident that he was 'quite disgusted with the RAF's smarmy ways', the way RAF personnel tended to duck out of the boisterous events often organized by the Army, the way RAF personnel often called superior officers by their first names and the way they wore fashion-conscious metal spectacles.

Local civilians tend to regard these 'facts' rather more favourably. 'The RAF are almost always well turned out, and they have the look of administrators or clerks about them,' is a comment sometimes heard. 'They have a touch of sensitivity about them, which you may or may not be able to say about Army squaddies,' is another.

Like the Royal Navy, which invites the locals to swarm all over their ships on organized open days, the RAF makes attempts to interest local people in its work. It organizes security open days with fierce Alsatian dogs chasing after men with heavily padded arms. These tend to be a great favourite with the locals – especially when one of the dogs lets itself be patted by children. Looking over the Harriers was a favourite local recreation. It has continued to be so with the Phantoms, against which it is popular to pose for photographs. Civilians sometimes give RAF men lifts in their cars, and RAF helicopters have been known to return the compliment.

But the main reason the local citizens are so sympathetic to British airmen is that RAF and Royal Navy planes were, for many, the first sign that Britain had not forsaken them. During the brief Argentinian occupation well-behaved Argentinian officers speaking excellent English told the Falkland Islanders that the British were just putting up a show and would never ever arrive to re-take the Falklands. While one young Argentinian officer, who had previously politely compelled a Falklander to run him and his men back to camp in a Land Rover, was waxing eloquently on this theme, two Harriers roared low overhead. The Argentinian officer did not show himself again.

For many, the Harrier was a signal that freedom was on the way. One Falklands resident, Richard Stevens, who farms with his wife Toni, recalled: 'The local population had a great affinity for the Harrier because, for many of us, it was the first contact we had with the liberating British forces. We saw the Harriers bomb or strafe the Argentinian positions in the saddle between Mount Kent and Mount Estancia. It was our first sighting of British forces, and to see the smoke billowing up from the Argentinians was fantastic. We didn't consider the loss of human life until later. A few days after this raid, a Harrier flew past us while we were out walking. We went berserk, waving and shouting, and it came past, low and close, and dipped its wings.'

Local citizens as a whole are aware of the RAF contribution in the Falklands campaign – and that lives are still occasionally being lost. One Harrier hit the ground near Walker Creek, and a Phantom hit Mount Osborne. At least two Chinook helicopters have also come to grief, one hitting the side of a mountain in bad weather (from which some of the crew survived) and the other crashing near Mount Pleasant, with no survivors. Local people attended the church funeral services in strength and contributed to funds set up for the bereaved families.

With civilians, the Phantom fighters are still reassuring but not perhaps quite as soothing as the Harriers were in their heyday. They are noisier. During low flying, if local folklore is to be believed, chimneys have been effectively swept, sliding doors have been ripped off their runners, and domestic ornaments have finished up in pieces on the floor. But surely, despite such minor irritations, the RAF cannot ever feel *more* sure it is among appreciative friends than when it is exercising around the Falkland Islands.

Possible overseas postings for the RAF tend to grow smaller and smaller in number as the years go by. Some take on, increasingly, the air of policing operations over what are ceasing to be strictly British affairs. In Belize, in Central

America, the RAF defends the territory of an independent nation against attack from neighbouring Guatemala, and has Harriers on stand-by to do the job. The Harriers would make short work of Guatemalan attackers, and everyone knows it, including the Guatemalans and the RAF.

In Hong Kong, too, the task can assume slight tinges of unreality, though it is regarded as a highly desirable posting, combining the tropical warmth of Belize with city amenities not available in the little fragment of Central America facing the Caribbean. There are dignified bars and dining rooms overlooking the old junks and modern freighters in Hong Kong harbour: they are high up in the combined services headquarters block, called HMS *Tamar* – the first custom-built combined services headquarters, and possibly, with the exception of the Falklands, the last. Within easy reach, there is a race track, a yacht club and the ferry-boats to the surrounding islands or to Macao, the Portuguese colony on the south coast of China to the west of Hong Kong, where there are casinos for those who want to gamble and antique shops for those who want to browse.

But the job itself has become inevitably more mundane for the ten two-man crews of the eight Wessex helicopters based in the New Territories of Hong Kong, the area nearest the Chinese border. The flow of illegal immigrants from the Chinese Republic has diminished to a trickle, compared with the heady days of 1979, when thousands tried to cross in a single day.

One non-commissioned RAF air loadmaster who spent three years in Hong Kong, from 1983 to 1986, told me he had encountered illegal immigrants only once – and that was by accident. He was on a training exercise with the British Army and was about to drop down on to a mountain side in the Wessex when he accidentally saw seven illegal immigrants directly under the aircraft. They were detained and sent back to the Republic. It was not, admitted the air loadmaster, a lot of excitement when spread over three years. Nor had the coming hand-over heightened the atmosphere of drama in one of Britain's few remaining colonial possessions; after the initial flurry when the British government signed on the dotted line to say they would hand over Hong Kong island, plus Kowloon and the New Territories on the Chinese mainland, both the local population and the British seemed to carry on much as usual. It is true that the RAF base at Sek Kong, in the New Territories, is virtually a training camp, not an operational base of the RAF in the sense that UK bases or those in Germany or the Falklands would be; the RAF is merely aiding the civil power in any way it can until its presumed withdrawal in the wake of the 1997 return of the territory to China.

In the Falklands, there is a constant threat – and many millions of pounds' worth of equipment and many hundred RAF professionals to meet it into the indefinite future.

In Belize, too, the future is uncertain. But there is little doubt that for the average RAF man or woman, the small Central American country is a more popular posting, with strong flavours of the sort of steamy, sunny and exotic places where the Union Jack once proclaimed participation in the British Empire.

Indeed, since Hong Kong came to resemble a Far Eastern model of New York, Belize is the posting which (with its tall jungles, primitive wood and straw shacks and open drains) most evokes the sort of atmosphere in which the people of the British forces had to live in the days of Kipling, when the empire embraced two-fifths of the world. It now does so for three hundred RAF men and women, twenty-two of them officers, who make up the regular RAF presence, armed with four vertical-take-off Harriers and four Puma helicopters, and enjoying a life which could hardly be more different from life in the Falklands.

If, in some seasons, the Falklands can seem like the north of Scotland, life in Belize is tropical throughout the year. In social atmosphere it is unique.

Belize is a curious little country of 170,000 people just south of Mexico, staring out towards the Caribbean. In atmosphere it is far more like Jamaica than Mexico or any of the other countries of South America, with their Spanish or other Latin history. It was once part of the British Empire, then won its independence. The red cast-iron pillar boxes still remain, though most of the cars are American and most of the hoardings advertise Coca-Cola and Seven-Up.

Impenetrable jungles divide the steamy south from the only slightly less humid north, so that what is now part of the Commonwealth often seems (except for those able to fly) not one small country but two even smaller ones.

The country's identity is not merely quaint. Its eastern coast is not far off Cuba. This makes North Americans sensitive as well as the Cubans themselves – since it is also near Florida. Behind it lies Guatemala, which has strong links with the USA, a stronger economy and military and a wish for greater control over the coast by which an invasion from the east could be mounted – Belize territory. It was threatening noises from Guatemala which caused Britain, which is responsible for the defence of Belize, to send out the RAF Harrier force and British troops in the 1970s. A force has stayed out there ever since. In the late 1980s it looked as if pressure by Guatemala for further negotiations was unlikely to come about until well after the next general election, not expected until the 1990s. Whatever the result of such negotiations, the fact will remain that the people of Belize are fiercely proud of their new independence and disinclined to part with it to their powerful neighbour or to anyone else, though they are prepared to accept help from outside. The open sewers in the main streets of Belize were due to be replaced by mains drainage and mains water under a Canadian scheme – not before time.

Part of the job of the Harriers (flown by pilots who serve for shorter periods than the usual six months of a Belize tour) is simply to reassure the Belize people that they are still there. When they first arrived, the local people danced in the streets, knowing that the Guatemalans had nothing remotely able to stand up to the Harrier. Now the RAF does between four and six sorties a day, trying to show themselves to every part of the country at least two or three times a week.

Care is taken not to irritate or provoke the Guatemalans. But pilots of the Harriers take photographs on a random basis at least twice a week to cover any activities on the border or in the air. They do this with automatic cameras to the left of the nose of the Harriers, which are themselves kept in secure conditions just off the main runway of Belize international airport. The airport, though also used by civilian aircraft, is nevertheless near military headquarters at Airport Camp.

One senior officer in Belize told me: 'It provides intelligence for us – anything of military significance about Guatemalan helicopters or troops – but may be of less importance than the more covert things, like the movement of drugs and the arrival of other illicit supplies.'

The other side of the RAF's activities is supporting the troops of the British Army, who spend much of their time hacking their way through dense jungle or wading through rivers infested with crocodiles and poisonous snakes. It often produces more tangible results. The Army suffers because maps quickly become out of date as whole stretches of jungle are burned down to make way for cultivation: what is shown as safe cover on the map may now be a flat charred terrain or a field of growing crops. The RAF goes out and photographs the terrain the day before Army exercises. These photographs, joined together, are used instead of maps by the officers briefing men for the exercise. In this way, the men have the advantage of information on the ground they are to cover that is only twenty-four hours old rather than, say, twelve years in the case of some maps. Most of the major destruction of jungle has taken place in the south, in the country between San Antonio and Jalacte, on the border with Guatemala. But even in other areas the terrain can change because of seasonal factors; and therefore up-to-date photographs supplied by the RAF are of more value than maps.

Harrier pilots also photograph aircraft which stray off course. They work in conjunction with the Royal Navy, which has a frigate in the Caribbean, but their main joint training is with the Army. During battle group exercises, all three services may work with the Belize Defence Force, a militia put together by the local population.

When I was in Belize a few years ago researching my book on the British Army, the Belize Defence Force was a rather gauche collection of local lads. RAF men say the standard has gone up somewhat in recent years.

'It is similar in strength and organization to a British infantry battalion,' one officer told me. 'I would say their present state is . . . reasonable. They lack decent equipment, although it is improving. They have two Defender aircraft – the beefed-up version of the civilian two-propeller Islander aircraft. They are supposed to be autonomous but we do help them with advice, with training and, I don't know whether I should say this, we give them a lot of equipment as well. No, not a lot, but we give them *some* equipment which would normally be scrapped in our service. Not many aircraft spares, but things like parachutes, which they use for dropping supplies.'

The Belize Defence Force has a loan service training team from the British forces, which includes two RAF officers and an RAF Engineer.

Twice a year in the Force field training exercise, the whole of British Forces Belize and the Belize Defence Force have artillery practice; they will fire missiles and drop bombs. And they will do it in a way they would not find easy over the cultivated countryside of Britain or Europe, since no one worries what they do above the jungle or out to sea in the battle ranges miles from anywhere. There they can happily drop up to one-thousand-pound bombs compared with the small cluster bombs they would be allowed to drop over West Germany (from where men serving in Belize have normally come).

There is a coastal range in the south of Belize as well as Mountain Pine Ridge in the middle of Belize. All three services of Britain can use live weaponry: the Royal Navy fires its guns in support of the Army light gunners, and the Harriers of the RAF weigh in with bombs, rockets and cannon.

The four Puma helicopters are largely there to carry troops, dropping them where they want to be for their own training exercises in the jungle and then resupplying them. This they do by dropping food and supplies into clearings in the jungle which are sometimes too small for a helicopter to make a safe landing. Twice a week helicopters go down south to practise with the troops of the British Army, but their role is not only military. There is always a helicopter on stand-by to deal not only with military emergencies but also with civilian ones. Usually they take the form of having to remove a Belize citizen from the middle of the jungle for medical treatment. Procedures here benefit from the fact that there is plenty of practice in dealing with mishaps to soldiers.

The Air Commander British Forces Belize, a Wing Commander with hawk-like eyes and experience in Bahrain, in RAF Germany and in the Joint Services Defence College at Greenwich, told me that he had recently written out a recommendation for a Queen's Commendation for one Puma pilot.

The pilot had been detailed to go to a small clearing in the middle of the Atap trees of the jungle – some of them over a hundred feet high and liable to sway violently in the breeze – and to rescue a soldier who had sustained a broken back when a tree had fallen on him. Getting neatly into the only available clearing would have been difficult enough at any time of day. But darkness was already falling when the Puma pilot arrived, his aircraft rapidly running out of fuel. If he stayed to get the soldier out, he might run out of fuel completely, or he might smash the rotor arms into the jungle around the clearing – quite easily, in the failing light. Yet the pilot knew that if they left the soldier on the ground all night he would almost certainly be dead by daylight next morning.

'It was a piece of magnificent flying,' said the Wing Commander. 'The soldier had fallen to the bottom of a dip. It was getting the soldier up the hill and into the helicopter which was very difficult. The pilot had to manoeuvre his helicopter under the jungle canopy to do it. It was dark by the time they had finished.'

Such expertise is often put to the service of Belize citizens. Sometimes local families or gangs fight with machetes, the long heavy knives used for cutting cane or clearing a way through the jungle. The injuries inflicted can be formidable. On one occasion, four men had to be flown by helicopter from Placentia on the coast to the Belize City Hospital, in the capital on the east coast. So much blood spilled over the floor of the helicopter that there were frantic mopping-up operations (in a different sense from the usual military phrase) before the RAF could pick up some British Army top brass they were due to carry.

One night, near Holdfast Camp on the western highway near San Ignacio, some soldiers were having dinner with a Belize family when a man burst into the house and shot one of the Belize men. The shot man was the nephew of the head of the family. It was thought that the attacker was from a rival family, and was intending to shoot the uncle. RAF men who got the shot man out suspected that the row had been about trade in marijuana or even cocaine. The troops took the young victim out to the airstrip, where a doctor was brought in by helicopter to stabilize the wounded man before he was taken to Belize City Hospital.

The joint forces medical unit at Airport Camp itself is, by general consent, the best medical facility in what is still a very poor country: the treatment I myself received there for a grossly swollen ankle (courtesy some bites by a few of the nasty creatures which abound) I shall always remember as being rather swifter than you would expect at home under the National Health Service. The unit is run by a British Army Lieutenant Colonel and employs a staff of about twenty to care for the one thousand five hundred or so British servicemen and servicewomen there at any one time.

And there *are* women of the Women's Royal Air Force in Belize, though usually no more than half a dozen at any one time among an officers' mess accommodating eighty.

Only officers are allowed to go to the country, where the temperatures soar to over a hundred and the hurricanes can be extremely fierce. But this, I was told, was *not* because of either the heat or the hurricanes. It was simply because though the accommodation for female officers was adequate, the quarters for other ranks was 'just too basic'.

Of the unmarried people stationed in Belize, 80 per cent are volunteers. They are lured, no doubt, by the prospect of sun and a change of scene. The RAF is realistic enough to know that, if it waited for volunteers from *married* people to fill all the necessary posts, it might be waiting a long time. Separation from family for six months is *not* popular; and even those who might welcome the thought of separation can hardly declare their preference publicly.

'They don't want to go, but it is like everything else in the RAF – someone has got to go and they all take their turn,' I was told by an officer. 'I suppose the comparison that everyone makes is with the Falkland Islands. In terms of recreation and general interest, I can only say Belize knocks the hell out of the Falkland Islands. Most units work a standard-length European working day. In Belize there are all sorts of sport, there is a cinema at Airport Camp, and other camps have their own cinemas as well. That takes care of mid-week. At the weekends they can do water sports and get out to the keys off the coast, which are a magnificent attraction.' They certainly are; they constitute the second largest barrier reef in the world, being beaten only by those in Australia.

Entertainment in Belize City is limited to a number of rather pleasant restaurants specializing in local seafood and a couple of brothels which are an incitement to celibacy. The state capital, Belmapan, is a series of concrete-built shops and a concrete-built cinema placed in the middle of the country (the middle of *nowhere*) to keep it clear of the floods which badly hit Belize City a couple of decades ago, an indignity to the members of the government not to be repeated. But Belmapan still has a population of only 1,500. It is not a Mecca of sophistication or culture. Very few diplomatic missions have moved to it; much of the real business of international relations and government is still done in Belize City.

I put to the Air Commander the view I had heard expressed that the British Ministry of Defence no longer saw Belize as serving strictly British interests and would like to withdraw in due course.

'I think withdrawal now would be extremely foolhardy, because the bottom line is that Guatemala has a territorial claim on the whole of Belize,' he replied. 'They have been trying to get some negotiations going for some time, but right now they are not going to settle it in a hurry because both the governments of Belize and Guatemala are getting near the end of their term.'

Certainly there is not in Belize, as there is for example in Cyprus, a significant number of voters who want Britain out. It has been estimated that the presence of British forces in the country gives it trade worth between 10 to 15 per cent of the gross national product. Those nice seafood restaurants would certainly have many vacant tables if it were not for servicemen.

Rather than encourage a British withdrawal, it would seem to make sense for Belize, even after reaching a possible understanding with Guatemala, to encourage British forces to use Belize as a training ground for things like jungle warfare. This would bring revenue to Belize, maintain a presence of British forces on Belize soil and give the RAF a continuing opportunity to practise low flying, which causes so many problems in Europe.

Belize is already a useful training ground for many skills, including survival in difficult conditions. The RAF Regiment do such training most regularly, but aircrews who may have to abandon their aircraft are taught how to make what is called a 'basher' (a palm-tree shelter), how to light fires and how to get water. There is less urgent attention given to the matter of how to find food, since aircrew are told that they can easily go for two weeks without food, if need be. If they are told it by the SAS, who run some of the survival training, they may perhaps more surely believe it – but they are taught all about how to trap and kill animals, just in case they get peckish before the fortnight is up.

Whatever the future may bring, the pattern of life for the RAF man and woman in Belize is today fairly stable. There is less chance of being stabbed or mugged than there would be, for example, in the case of British people in some countries of the Caribbean proper. Ten RAF officers or senior NCOs are able to be accompanied by their wives and children; and they claim that education is perfectly satisfactory in Belize up to the age of eleven, though after that it probably means a boarding school in Britain.

Officers' wives tend to be an animated little group, doing keep-fit classes (no mean chore in the Belize heat) or visiting the local caves and keys. Coffee mornings are de rigueur.

There are definitely shades of the Kiplingesque about it. The most *indefinite* thing about it is how long it is likely to *go on* as a way of life for Kipling's heirs. The consensus among the men and women in light blue seems to be that it would be a bold man who would make predictions more than five or ten years ahead. They have the compensation of knowing that, at least for now, they have rather less tough lives than some in the Royal Air Force.

5

The One and Only Regiment

'My wife said to me when she knew I was being sent to the Falklands, "I don't want to see you on television with your hands up. I would rather you died – I will be comforted and well looked after if you die, don't you dare let me see you surrendering." I told her not to worry.'

– Officer of the Royal Air Force Regiment.

'There is no closer link in any of Her Majesty's Forces than that between the RAF Regiment and the United States Air Force. We have got the most intensive co-operation. Perhaps there are nasties who would make something of it, but it really *is* a remarkable co-operation.'

– Director, RAF Regiment (an Air Commodore).

The gung-ho cinematic antics of John Wayne have not necessarily made the United States Green Berets the most beloved military force in the world. Even the Red Berets of the British Army Parachute Regiment have an image that is closer to the SAS than, say, the friendly village bobby. But the Blue Berets of the Royal Air Force Regiment (and only partly because their job of guarding RAF installations is a *defensive* one) can sometimes have a soothing effect on touchy situations: these are ones where the Blue Berets are able to appear more like peaceful policemen than high-profile soldiers.

The RAF Regiment consists of three hundred officers and three thousand other ranks, whose job it is to protect airfields and equipment from enemy or subversive attack all over the world. They form the only regiment in the RAF.

It has been fashionable in some circles to poke fun at the regimental system, on which the British Army is based, characterizing it as a fuddy-duddy remnant of feudalism, in which the officers are the gentlemen, the other ranks are the peasantry, and everyone Knows Their Place. It might be truer to say that, in many respects, the regimental system is a *less* snobbish one than a system based on an officer corps which exists as an entity across an entire service, owing a loyalty to itself rather than to a unit embracing *all* ranks like the regiment – which is seen by its defenders as more like a family, with its rich and poor and its young and old, rather than as a more coldly drawn 'organization' structure.

Officers of the RAF Regiment are convinced that the regimental idea is its strength. They believe that, no less than the British Army, it is a tough soldierly force which is, however, no stranger to diplomatic niceties. They argue that it has certainly proved itself in those terms in sensitive operational areas.

The Director of the Regiment, a spry Air Commodore, told me that the Regiment has at times been used not merely for the immediate defence of RAF assets. 'It has also been used when there has been some hostility to other forces – because emotions can simmer down once the Blue Berets appear. We assume a low profile, and it seems that it cools passions. This applied during some of the troubles in Cyprus.'

On such fine points do defence forces sometimes rely in Britain's post-imperial area. But in no other sense can life in the RAF Regiment be seen as soothing. It is their presence that still guarantees British air presence, not only in Northern Ireland and Cyprus but also in Germany, the Falklands, Belize and Britain itself. And the fact they are at their depot at RAF Catterick in North Yorkshire at breakfast does not necessarily mean they won't be in Cyprus by nightfall: their job is unpredictable.

About the only predictable thing in their lives is that, at any one time, they will be on active operations *somewhere*. Since they were formed at the time of the Second World War, when airfields became of acknowledged crucial importance, there has been only one short time (eighteen months around 1959–60) when they have *not* been on active operations somewhere in the world.

When Rhodesia declared its unilateral independence from Britain in 1965, RAF Regiment officers at Catterick were warned at eleven o'clock one morning to be ready to move. At this point they were not told where. At one in the afternoon, a squadron was put on immediate readiness to move. At eight that night, it left for Zambia. No one could tell his wife and family what was happening *before* he set off, because he didn't know. Much the same thing happened in 1963, when trouble erupted between Greek and Turkish Cypriots for the first time since Britain conceded Cyprus its independence – just as the RAF Regiment, when Cyprus was still under British control, helped to search for the guerrilla leader Grivas.

By the time, in 1977, that Guatemala seemed to be on the point of mounting an invasion of neighbouring Belize, senior officers had become used to the military-diplomatic drill. One officer recalled hearing a knock on his married quarter door at one o'clock in the morning. It was an orderly officer, who told him to be ready to move by eight that same morning. Number Three Wing took off for Belize shortly afterwards.

When the Falkland Islands were invaded by Argentinian forces in 1982, and the British task force was sent down to the South Atlantic to eject them, a unit armed with the Rapier missile system was needed. The only units on standby at that time were those of the RAF Regiment in Germany. A squadron was plucked out of Germany in eight hours and went on the *QE2* with the Fifth Brigade, landing at Goose Green.

The RAF Regiment squadron in the Falklands has in effect, in the view of some officers, become an established part of the training system of the RAF Regiment, and as such, a way of getting value for money out of the Falklands military presence. In 1982, one senior officer (who was to become Deputy Director some years later) encountered a rather less orderly situation. He was just starting a period of leave (or *thought* he was) when he was summoned and told he was being sent to Ascension Island, four thousand miles away from Britain, and the halfway house staging post to the Falkland Islands. It was feared that the Argentinian special forces would try to sabotage British aircraft at Ascension, by either seaborne or aerial attack. For a while, the RAF Regiment commanded both the Army and the Navy on ground defence, as the recently sent officer drew up a ground defence plan, writing it on an open-air 'desk' consisting of an egg-crate, in winds of up to forty-five miles an hour. Later he handled the transfer of Argentinian prisoners.

The officer looked on the bright side, in retrospect: 'For the first time, the RAF Regiment was involved in ship recognition and coastal defences. That was valuable extra experience.'

One of the reasons the RAF Regiment are provided with light armoured vehicles is that, in a European war, their transport would need to be splinter-proof. But the armoured vehicles would help them fend off attacks on any RAF installations at home or elsewhere in the world, helping provide *double* the amount of fire power that the same number of conventional infantry would provide. Twice the number of machine guns that conventional infantry would have assists their heavy fire power; and mobility permits the RAF Regiment to dominate an area of about 170 square kilometres around most airfields.

But if the fire power is relatively massive, the structure of the RAF Regiment followed the principle of 'small is beautiful' long before the phrase was coined. From the start in 1942, the primary unit was the squadron, consisting of perhaps 120 to 160 men. Backed by their Royal Warrant issued by King George VI, establishing them as a Corps within the Royal Air Force, the RAF Regiment men in these small units were more or less self-contained, with their own badges, their own chain of command, their own medical officers. Their structure, in this respect, is akin to that of the Royal Artillery. Under this system, a young officer can find himself responsible for thirty-three men, including, possibly, a Flight Sergeant twice his age.

'That's where we get our strengths – young pilots don't have charge of men in that way,' say veteran RAF Regiment officers. 'It rapidly develops fellowship.'

It would need to. By the late 1980s, some officers and men had already experienced as many as twenty unaccompanied four-month detachments in the Falklands and elsewhere, creating inevitable problems of separation for their families. It is here that the essential regimental structure comes to the families' aid. 'Shadow' squadrons of wives at home deal with many of the problems arising in their early stages, well before they escalate to official levels.

While cynics might argue that the British have to go to the Japanese to find out how to build and sell a motor car, other countries appear to be still eager to model their fighting forces on British models and skills.

Even the Americans called on the RAF Regiment for advice when they realized they were finding it difficult to defend their air installations against guerrilla and other attacks in Vietnam. Co-operation with the Americans goes beyond the units of the RAF Regiment which are assigned especially for the protection of US Air Force bases in Britain, always officially designated as RAF bases. It was the RAF Regiment that advised the Americans on the whole concept of Combat Security Police, which the Americans were to send to Vietnam and other areas of conflict. An integrated RAF–USAF force was set up at Molesworth and at Greenham Common, where the intermediate-range nuclear missiles were kept.

One RAF Regiment officer told me: 'In consequence of this terrifically close liaison we really do have an immensely close relationship, as if we were part of the same force. They watch our promotion list and we write letters of congratulation to one another. Many of us spend holidays with US personnel. We have inter-service competitions with the Americans.' Students of Napoleon's victory at Austerlitz, over a *squabbling* alliance of Austrians and Russians, will find this accord encouraging.

In continental Europe, there is a close interest in the way the RAF Regiment works. The French have a force directly comparable to the RAF Regiment. The Belgians also established a force directly modelled on the RAF Regiment in 1945, after the Liberation at the end of the Second World War: their present technique is to select an élite from their Air Force as a whole, and place them in the ground protection force. The Swedes, though non-aligned in peace and neutral in war (at least, that is their *hope*), have lately shown an interest in the methods of the RAF Regiment, though their thinking about preserving their air force appears to be based on the theory of dispersal into a hinterland in which they can hide, rather than on defending established airfields. The Swedes wanted to know how best to detect any threatening movements towards their forces, and how best to use guard dogs in helping to safeguard installations.

Japan recently asked for advice about air-base defence. Spain has shown an interest in the use of the Rapier missile, and Turkey has asked for advice on training. A number of Third World and Commonwealth countries have sought advice. In the early 1970s, the Argentinians sought advice on one defence system.

The Director of the RAF Regiment had no doubt about it: 'We are world leaders. It's been a very successful model, the RAF Regiment. Despite all the turbulence of the defence cuts, the RAF Regiment, with the Parachute Regiment, the Royal Electrical and Mechanical Engineers and the SAS, is one of the few organizations formed in the Second World War that have survived until today. The bitter experiences of the Second World War showed that you can't operate without it.'

With one reservation, Germany is a popular posting for the RAF Regiment. The reservation, as I observed when I visited 37 Squadron at RAF Brüggen, among the cluster of RAF stations in West Germany, was the fact that once every sixteen months, the squadron had to spend four and a half months in the Falklands, whereas if they were normally posted in the United Kingdom, they would spend this time in Belize.

Belize was far more enjoyable, one twenty-nine-year-old Corporal told me. You could have a bit of fun in the sun – it was more like a holiday than a tour. You could go around and see the countryside. After eleven years with the Regiment, he liked the visits to the Falklands much less than those to Belize, Norway or the USA. 'Speaking personally, it is such a pointless task out there. For four months you are doing the same thing, day in, day out. You are on full vigilance in the Falklands, on a war basis, and I don't think people see the point of that. A few think we could get away with being laxer. Facilities have got better for other personnel, since the airport was moved from Port Stanley to Mount Pleasant, but really what we have done is move from one piece of grass to another just down the road. It is tedious and the boredom gets to you. People just become lethargic and wait for the plane to take them home. There is no let-down in discipline, because most are mature enough to do their job, but not as enthusiastically as they would do it here in Germany or anywhere else.'

In Germany, unlike the Falklands, members of the RAF Regiment can and do organize exercises which simulate the reality of the threat. In some respects, exercises are easier in Germany than they are in the United Kingdom. There is a device called a 443 Order, which immediately authorizes the use of land for military action. The men are 'blocked' into a particular area, and there are restrictions about standing crops; but it is still easy to train.

I asked a Flight Lieutenant of 37 Squadron, a thirty-five-year-old northerner, what relations with the German civilian population were like, and was assured that they were generally excellent. He could not remember any clashes, and it amazed him how the Germans were prepared to allow their land to be used for military exercises. He remembered an exercise at Gütersloh a few weeks previously, in which they had parked their Rapier missile unit in a farmer's back yard. They waited for a stream of invective when the farmer first approached them. There was no invective. Instead, the farmer pointed out where the tap was, in case they needed water.

Perhaps the fact that farmers are compensated for damaged crops (and have even been known to beckon military men into fields where crops are bad, knowing that the compensation will be worth more than the crop) helps cement good relations. But the basic reason is simply that the Germans, having been invaded themselves, are much more aware of the value of the RAF Regiment.

Much of the RAF Regiment's time in Germany is spent living rough, under canvas. This means living *very* rough, under what is called a poncho – a big plastic square which is hung from trees, a couple of feet above the ground. The men wriggle under it and use their kit as a pillow and wind-breaker. The Regiment usually spends three days a month living like this, sometimes when temperatures drop to as low as twenty degrees below freezing – as cold as it gets in the Arctic Circle. Ordinarily men will sit with the Rapier missile launcher for a two-hour spell, but this is reduced to half an hour in such cold, after which the man is put into a tent with a heater until he thaws out. But unlike exercises in Norway, no one in Germany has to sleep in snow-holes. The snow is not deep enough at the levels the RAF Regiment have to work at – the level of the airfields – though it may be at the tops of mountains.

During exercises in Germany, Rapier teams will be virtually under the command of the Sergeant on the spot, as they would be in war. A Rapier team is commanded by a Sergeant, who has his strict firing orders; but if the group become cut off from communications to the control post, the Sergeant has to interpret his orders himself, and give his own firing instructions against an enemy menacing the airfield. In exercises, this is done on a twelve-hour shift basis. Officers comment drily that they prefer a twelve-hour shift in the field to twelve hours in the command post with the Commander in Chief.

At both work and play, discipline is retained in a way more reminiscent of the British Army or the Royal Marines than the rest of the RAF. Officers and men say that relations between officers and men are *easier* because a formal front is always maintained. Senior NCOs formally entertain officers in their messes and vice versa; but address is always formal – an officer calls a man by his surname, and the senior NCO or airman calls the officer 'sir' at all times.

'It is good that there is no difference in the way I address the troops at eight o'clock at night and eight o'clock in the morning,' said a Flight Lieutenant. 'Blokes know where they stand. It is not Jim and Fred one moment and Smith and Bloggs at another. That wouldn't work.'

Even in the conditions of annual training camp, for members of the RAF Regiment doing a full professional job, the rigid rules are not relaxed. Such camps are, however, generally popular with active squadrons of the Regiment, who can eventually tire of guarding the same old airfield. I certainly found this to be so in Cyprus where No. 34 Squadron had broken off its guarding of Akrotiri airfield on the south coast to travel seventy miles north-east towards the border of Turkish-occupied Cyprus. There they were training within the second British Sovereign Base in Cyprus, that at Dhekelia.

Another light-armoured squadron had moved into Akrotiri to free No. 34 Squadron for their four-week camp. In a valley, twenty-eight tents were set up for the one hundred and fifty men and seven officers. The Commanding Officer had a twelve-by-twelve foot tent to himself. The three Flight Commanders shared a tent the same size, and the airmen – also in the same-sized protection from the weather – shared six to ten to a tent. No pretence at equality here? Officers point out that the Squadron Leader's tent is also his HQ and that he holds regular training and other meetings in it. In any case, the standard field kitchen fed everyone, at the standard allowance of ninety-five pence a day.

Too much was going on for any dissatisfaction to build up. It involved the whole squadron as an entity, which no training at Akrotiri airfield itself could ever do. At the Pyla Range, an old quarry with its neck pointing out to sea, men fired two-inch mortar bombs and sixty-six-millimetre anti-tank weapons and threw grenades in stalking exercises. They used machine guns. Each man fired at least three two-inch mortar bombs at targets placed at three hundred to four hundred metres. During anti-tank stalks, each man fired twenty rounds, which in war would pick off enemy soldiers trying to escape from a crippled tank. On night shoots, riflemen fired eighty rounds, and handlers of the general-purpose machine gun had the chance to fire two hundred rounds. This was while section commanders fired six rocket flares for illumination.

Only those familiar with the cost-cutting measures that have operated in the services today will realize how liberated members of the RAF Regiment feel when they are suddenly given access to all this ammunition. The Deputy Squadron Commander, a steely eyed Scotsman with a broom of a ginger moustache and a chin that could have been used as a battering ram, surprised me by saying that, yes, he thought they *did* have enough ammunition for proper training.

'The restrictions we have to place on them are because of the amount of *time* we can spend on the range. That is the problem, rather than a shortage of ammunition. I have got thousands and thousands of rounds back at Akrotiri.' The only exception was with the high-explosive tank sixty-six-millimetre rounds, for which twenty-one-millimetre-calibre rounds were sometimes substituted.

I felt I could take the satisfaction with the ammunition supply situation at face value since, about other aspects of their training, the RAF Regiment officers were obviously not inhibited about voicing any doubts.

The Deputy Squadron Commander was frank. 'We have eight vehicles with us on which to maintain our skills. Normally a squadron would have twenty-three. The rest are back at Catterick headquarters, being worked on by the Light Armoured Training Squadron.' Grey eyes glinting, he fired off a few more verbal salvoes when I asked if training could be realistic in Cyprus for what, in war, would be a German campaign: 'The reason it isn't realistic is that it's a completely different landscape. We are training in a semi-desert, but we will be fighting on the North German Plain or wherever – including built-up areas. It comes down to the difficulty the RAF sometimes has in adapting to its environment. Look at my vehicles. Dark green, or green and black, and the same for my operational Land Rover. Everything in Akrotiri is light brown. If some Arab wants to shoot me, I stand out. There should be some light brown paint, but the RAF doesn't have any . . . The Ministry of Defence will say, "It is all part of some big plan." I don't believe in big plans, anyway . . .'

Senior officers reject such strictures as too sweeping or, as one of the top echelon put it: 'He is talking rubbish.'

Plainly, members of the RAF Regiment consider themselves blunt-talking soldiers, rather than the usual run-of-the-mill sophisticated men of the RAF. One of the Sergeants told me that there was a different pace of life in the RAF Regiment, one that many RAF men could not keep up with. 'When I was on basic training, I found that people who used to join the Regiment had a lot of qualifications. So they are not thick – but they don't want to be indoors, stuck behind a desk. They would rather be out of doors, with a Rapier missile, or with armour, out in the field. They take to camp much more quickly than others in the RAF, because they are training from day one to do out-in-the-field soldiering. In my personal opinion, the RAF are tradesmen before they are airmen, but we are airmen or soldiers first, and then specialists.'

This view is strongly disputed by the Regiment's senior officers; and the rest of the RAF would certainly dispute coming off second best in this comparison. But there is no doubt the RAF Regiment in Cyprus – especially in camp, with its canvas weighed down with rocks, gas cylinders or fire extinguishers – rather prides itself on its hardships. Training on the electronic pop-up target range, the moving target range and close-quarter battle range may have its satisfactions, but it was the bayonet practice area of which No. 34 Squadron seemed most proud. Officers admitted – or boasted – that it was a 'fiendish bit of kit', a human-sized dummy with a parrying arm which could strike the wielder of a bayonet, just as the enemy might do, if the attacker were not careful. The clincher was that a trainee had fractured a couple of ribs on it only a few days before. This, to his colleagues, seemed to amount to some sort of battle honours.

The Deputy Squadron Commander personally cut into this anecdote to play down, for the benefit of a civilian observer, the harsher side of RAF Regiment life: 'There are still advantages to Cyprus for us – it's a good family tour. Men in the United Kingdom may spend eight months away, which can be divisive in terms of the family. Here they will only be two to three months away.'

For the tough members of the RAF's one and only regiment, such benefits amount almost to flamboyant luxury: only those 'other men' in pale blue, compared with their khaki, might *expect* such things as a matter of routine and right – or so you can sense these men thinking.

Other *men*? One should hastily correct this to men *and women*.

6

WOMEN AND WIVES

'It would be a good idea to allow women to become pilots. The Americans are doing it. The Dutch have just failed their first would-be woman pilot. The RAF are twenty years out of date, and I can't see it happening in our lifetimes. I will believe it when I see it.'

– Senior aircraftwoman stationed at RAF Brüggen, West Germany.

'Women are more integrated into units now. You simply can't afford, given the manpower shortage, to use women as a sort of special category who can be used only for a limited number of jobs. A lot of women here are doing what ten years ago would in all probability have been an exclusively male job.'

–Warrant Officer at communications centre, Episkopi headquarters, RAF Cyprus.

The senior aircraftwoman had just been listening to five electronic conversations at once to get the information necessary to make possible the landing of half a squadron of Tornado aircraft. Petite, twenty-five and with a crisp and guarded manner, she said her job in the control tower of the airfield was 90 per cent boredom and 10 per cent panic – and you had to know *exactly* what you were doing in that crucial 10 per cent of the time.

She moved away from her seat at the control tower of RAF Brüggen, a station among the cluster of RAF airfields in West Germany, and the home of the Tornado strike-attack aircraft with their nuclear capability. It was only then that she beamed her vigilant attention directly at me.

'We don't actually control the aircraft, but we take everything else off the backs of the traffic controllers,' she said. 'When everything starts piling up, you have to be on the ball. You do listen to five conversations at once, which is one thing people who are in the control tower are good at. We require information, and we get it from several sources, and we pass on whatever the air traffic controllers need to know. I prefer to be busy and pushed, because I work better that way.'

The RAF obviously has no blanket objection to responsible jobs being done by the youthful or by the female. This experienced senior aircraftwoman was both and she was also an enthusiast; ever since she was fourteen, she said, she had wanted to do either journalism or air traffic; and, since her parents were both in the RAF, the RAF won over the Fourth Estate.

Did she find she was treated differently because she was a woman? Some of the physical jobs were more difficult for her than the average man, she said. On the runway were about forty petrol drums, lit with batteries. During winter, each evening they had to bring these back to the battery bay for recharging. She found that quite hard, but had to do it the same as the men, and in the same amount of time. There were marking lamps filled with sand, which were heavy, and which a man, rather than her, might be sent to lift, if a man were available.

She thought some men considered that the women had it easy. If fact she thought it was more difficult for them: 'Because you have to get on with the men, which can be difficult if they think you are having an easy time, not having to do things they are expected to do. In general, we get along very well. If you couldn't treat them as one of the boys when you were out at work, you would never hack it. I have always worked more with men than with women – I was in the Department of Health and Social Security as a clerical officer before I entered the RAF – and I can cope all right.'

What did she find presented most difficulty? The fact of not being very tall. There were two Land Rovers they had to use, and one of them had such a high side it was difficult for her to hitch up her skirt and climb into the vehicle. But she had to wear a skirt except when driving in winter, when she would wear trousers. No, there was nothing actually laid down saying a skirt must be worn generally, but it was simply *expected* that you would do the job in a skirt. Even when dealing with, say, barbed wire? 'It is only in the winter that I would wear trousers,' she repeated firmly.

I asked the senior aircraftwoman whether she thought the RAF, on the whole, was a good employer of women. It seemed, from her answer, that the only difference between men and women at Brüggen was that the women, though trained in the use of weapons for personal defence, were not expected to augment the guards round the base at times of tighter security.

As a woman, how did she view the possibility of having to shoot someone? The answer was suitably laconic. 'If somebody stood there with a gun pointing at me, I would rather have one in my hand, saying, "I will have you first, sunshine!" They used to issue us with pickaxe handles before women became trained in personal arms – crazy!'

Women in the RAF do not necessarily get the same pay for the same job as the men. The senior aircraftwoman at Brüggen said this was not a sore point, though personally she didn't see why women should not be paid the same.

Perhaps a little clue was provided when I asked her what were the usual topics of conversation when women met off duty. By then I had been in numerous messes, where I soon discovered that work is almost invariably the subject of conversation among the men, suggesting an ultimate commitment to their craft. In contrast, women often appear to be just as conscientious at work, but more determined to maintain a separate private life.

'When I get out of work,' said the senior aircraftwoman, 'I don't speak about work. To me that's *it*. I get on with my social and personal life.'

She had just married an aircraftman technician, and she did not want to become an officer. Once she *had* wanted to be an officer, but without good mathematics it would be impossible, and now she wasn't worried. Her husband didn't want to become an officer, either, and she was quite happy as things were.

Perhaps because women *expect* the RAF to be more progressive than the other services in its treatment of women, their discontent, if they are disillusioned, can be the greater. This is especially discernible when they are in an overseas posting, where social life may be restricted and grievances can become magnified. A group of aircraftwomen I spoke to in Cyprus were unstinting in their praise of the good things about the RAF – particularly its variety and sporting facilities – but also voiced a number of reservations.

All of these could perhaps have been summarized under one heading, and one not unknown in civilian employment: Not Being Taken Seriously Enough As a Responsible Person. They said the accommodation was confined, with many of them having to share a room: but neither this nor the dustiness of the accommodation in the Cypriot climate was the thing that really bothered them. One of them said: 'They still think that single girls live out of a suitcase, and they don't.' But this was not the major cause of complaint.

The chief cause was that the aircraftwomen felt they could not behave as they liked in their residential blocks. Though officers could have alcohol and men in their rooms, other ranks were forbidden both. This obviously annoyed them considerably, though the official line is that their view is a misconception.

'You are expected to be adult, yet you are told by the rules of your accommodation that you are a child,' said one. 'If you want to bring a man in for a cup of coffee you can't do it, and if you are going out with the girls and want to have a glass of wine before you set off, you can't do that, either. It is not right. Officers and senior NCOs are not restricted like this, and yet some corporals can be in their thirties and some sergeants in their twenties. It is all wrong.' (The official line is that a different-sex friend *can* be put up if a visitor's room is available).

All the girls – a group of eleven – were quite comfortable with the fact that officers and senior NCOs did not share their mess. They pointed out that it would be difficult to exert the right discipline as an officer if you had been drinking the night before in the same mess as the aircraftwomen. What they were *not* comfortable with was what they saw as two distinct sets of rules – one for officers and senior NCOs and one for others.

'In the Naafi you can go and get drunk, and if you damage anything you can be charged,' said one. 'The same rules should apply in your accommodation. If you were allowed alcohol and men in the block, there might be a great fuss for a fortnight; after that it would settle down and the result would be about the same as now – when some people *do* drink despite the rules.'

One well-built and obviously capable girl said that, at one stage of her RAF career, she had been in a split-sexes block, with women on one floor and men on another. 'It was the best block I have ever been in,' she said. 'People behaved in a more civilized way, whereas now the men sometimes get drunk and smash the place up; and we did not get things pilfered from our rooms because the men were at hand.'

Others said it was ridiculous that a man could be brought only into the sitting room, where other girls would be watching television: 'It is not worthwhile, it is embarrassing, and soon you stop asking them in for a cup of coffee.'

The women blamed the men for a prudish attitude about the presence of women in some close working conditions. A sturdy lass who looked capable of stunning man or mule with one blow of the fist said she had been on joint services training in camp, sharing a tent with seven men.

'One of the men said, "You can't sleep in here with us." I said, "If you don't like it, you can complain officially." They were extremely prudie. If you get seven men and one girl they will be prudie. They were basically a good crowd, but two guys, married guys, winced as I took off my shirt and asked what I was doing. I said I was washing, something that women had to do as well as men. Fortunately I was *in charge*. I slept in the tent.'

The girls claimed that the basic rule was that women could accompany men in working conditions only if they had their own toilet facilities – but said that the rule seemed to be interpreted in different ways. One accused the RAF of double standards: the RAF allowed women to doss down with men when it suited them, but became prudish when it didn't. She said she had been in the operations centre at Episkopi joint services headquarters at Cyprus, but had been moved to another job because that one meant doing a twelve-hour night shift alone in an office with a man; a man moreover, who was an *officer*. 'People do it always at night, never in the daytime, you see!' she commented sarcastically. 'At first they brought in a corporal to make it a threesome; so I suppose the officer and I could have got on with it while the corporal did all the work. I couldn't believe it when I was given a male chaperon. I was very annoyed about it.' At the time of our conversation, she had been transferred to the air traffic tower, where presumably the large square footage of window space was thought to make any sexual malpractice more easily detectable.

The girl with formidable biceps claimed that sexual harassment was prominent in the RAF, though the rest of the group seemed prepared to treat it as a rather poor male joke. 'Most people accept that it is a man's Air Force and a married man's Air Force at that,' she argued. 'If you are a female, you must do at least 50 per cent more than a male would have done. And if a guy makes a pass at you, and you go to a superior officer and say, "This guy keeps touching me up", he will be a *male* officer and he will say, "I can't prove it." The men can go on saying things like, "Nice pair of boobs you've got there, darling", but nine times out of ten your boss you go to is a male, too.'

Perhaps this particular girl got teased primarily because she was the sort of hearty person who could take it, as well as perhaps hand it out? She was certainly robust in her determination *not* to be what she considered the *men* considered to be a 'typical WRAF': a woman who winced or cried when required to pick up a twenty-five-gallon drum. 'I have to work very hard to convince them that I am not one of those who snivel when things go wrong.' (Officers point out that all stations where airwomen are employed have female officers and NCOs.)

Most girls were not so much concerned with the ideology of feminism as with the use of their skills in the RAF. It was certainly significant that, when I asked the group who intended to make the RAF a permanent career, as distinct from a stepping stone to marriage or some other job, nine put their hands up. It was perhaps equally significant that one of the other two was leaving because she had married an Army man and joint postings would be difficult, while the other one said she was considering leaving because she felt under-used.

She had trained as a secretary but instead of going ahead to be a personal assistant (her aim) she was in fact doing a junior clerk's job, sifting letters and doing the filing. 'I know it sounds snobbish, but I think it is a bit beneath me,' she said. 'I have put down for PA duties, and I think the RAF could give me a chance. I am not planning to get married at all.'

I asked her if she would leave the RAF if she married. 'I don't think it is fair to have to follow a man around,' was the immediate reply.

I was not surprised to hear that this girl had shown previous signs of uneasiness within the RAF. She seemed to be an intelligent girl who would probably do well in civilian office employment but was not necessarily a 'natural' for a career in the armed services. The RAF, like the other forces, must be easier for (a) career girls who love the military life as such and (b) sporty girls who will leave the force when they find a husband. The frustrated secretary fell between these stools and seemed to be getting the worst of all worlds.

But for the rest of the aircraftwomen, the revision of the accommodation rules against men and alcohol, a clear set of rules on the mixing of the sexes in working conditions and at least some further consideration of the possibility that women might become pilots would go a long way towards placating their doubts.

Not that the desire to be a pilot was anywhere near unanimous. One small, pertly spoken girl said that if women could carry arms, which they now could, they could also shape up to become pilots. But the robust lady who complained of sexual harassment surprisingly took a different line: 'The problem, if you want to be a fighter pilot, is that, with the forces operating on the body in climbing and turning and diving, your internal system would probably drop out. To be a fighter pilot, you would have to have a hysterectomy and, to me, it isn't worth the price.'

But she did not argue further when one of the other aircraftwomen said: 'Yes, but if a girl wants to take that risk, she should have the chance.' It was obviously an ideologically sore point to many of the aircraftwomen, even if they had no ambitions in that direction themselves.

Such irritations tend to multiply when the women are doing taxing work over long hours. This situation often applies in Cyprus. In the communications centre at Episkopi headquarters, there are twenty-six girls working in shifts, either from seven in the morning till seven at night or seven at night till seven in the morning, with 'sleeping days' built into the system for them to recover. Their business is to scrutinize secret messages coming in by tape, to evaluate them and pass them on to the right quarters. This sort of work is deprived of the normal working safety valve: the ability to grumble about it to people outside. Women in this work tend to marry men in the same trade, which at least makes conversation about work possible with them. Otherwise, it must be a pretty lonely furrow to plough, especially when miles from home in Cyprus.

One twenty-one-year-old senior aircraftwoman told me she had always been interested in this sort of work, but sometimes found 'talking to computers' less satisfying than talking to people. A resentful gleam came into her eye only when I asked her about men.

The answer was decidedly frosty: 'There are plenty of them here, but there is not much to say on that subject. They tend to treat you very unfair. Not so much the blokes you work with, because you get to know them. Generally blokes on the station as a whole tend to degrade you just because you are a female. They put you down in any way they can, by their comments about your figure and morals. The majority think you sleep around and that is all you are here for. I think it is a general service outlook. I suppose you get it in civilian life, but because here it is such a family atmosphere you notice it more. It only applies to a minority, and you tend to ignore it.'

The senior aircraftwoman had been in Cyprus only three weeks, which may have been partly responsible for her sensitivity. A twenty-four-year-old with three years' experience of being in Cyprus, on different duties in the communications centre, said that, on her free afternoons, she played netball, volleyball or hockey, or did some running. She was more optimistic about the attitudes to women in the RAF: 'I think it is getting a lot better. It used to be thought that a woman joined the RAF to find herself a husband, and would then leave. Gradually, women are staying in more, and more women are promoted. The previous attitude seems a lot more rare now. The attitude of men you actually work alongside is quite good really – they look upon you just as some person they work with.'

The irony was that this senior aircraftwoman had met and married her husband when she was training in the RAF and her husband was already in it. A further irony, at the time I spoke to her, was that she was due to be promoted; whereas her husband – the same rank as she was – was due *not* to be promoted.

This sort of situation must be the touchy side of Women's

Lib, posing a few problems for the people concerned and for the RAF as an organization. In this case, after the woman's promotion, both partners would come back to the United Kingdom and be at two separate camps, five miles apart. This would avoid the man having to defer to his Corporal wife, while keeping the married partners fairly close together. It was now, said the Corporal-to-be, quite common for wives to hold the senior rank.

What did her husband think of this prospect? 'He is pleased because it means more money. He doesn't want a career with the RAF anyway, whereas I think more on the career side. He wants to join the police – he wishes he had joined them in the first place. One of the reasons he is leaving the RAF is that promotion is rather slow.'

Perhaps one had to read between the lines of this answer, the more so since I had no opportunity to meet the husband. But the Corporal-to-be's attitude to the RAF was very positive, one of her few reservations being the standard uniform in Cyprus – a tightly fitting beige-coloured one-piece 'shop assistant's coat' as the Corporal-to-be put it (actually its very plainness emphasized her femininity; but of course it is the way the wearer *feels* about clothes, rather than the objective effect on other people, that really counts).

'It is shapeless and the material is rather thick for this climate,' she told me. 'There is no way in the system that I know of in which you can suggest a new uniform. They tried a blue version of this. It didn't go down well, because there was no design change.'

So do women in the RAF who *say* they want equality with men also want to have their cake and eat it, in being able to worry more than men about their appearance? Certainly not all of them. In the private branch exchange of the telephone system at Episkopi I found a twenty-one-year-old senior aircraftwoman, married to an RAF technician in the same trade. Her normal job, she told me, was servicing communications equipment. 'It involves a lot of work with the soldering iron,' she said. 'I have always been interested in getting my hands dirty. I did metal work at school. I like to think the men take me seriously as an engineer. There's a bit of banter, but they don't try to make you feel an outsider.'

Her ambition, she said, was to go on a fitter's course, involving eighteen months of largely theoretical work. At the moment she would be more likely to *assist* a fitter: 'When you see fitters doing complicated tasks you think, "I wish I could do that." At the moment, all the fitters are men. I will be posted somewhere else when I get promotion.'

This pattern of life seemed markedly different from the memories of an older colleague working alongside her in the telephone exchange, both of them near the plum-red 'hot line' receiver which both watched like hawks. The colleague was forty-five, married to a Flight Sergeant working in the communications centre and had vivid memories of her own youth.

Even when she was twenty-one, she remembered, she had to be in by eleven at night, whereas now that applied only until you were eighteen. 'At twenty-one I was being bed-checked – someone came round to check that you were in bed. It was only if they were Corporals or over that they could get chits to stay out until midnight. They have got a lot more freedom than they had in my time.'

They certainly have more freedom in their choice of work. Women crop up in all sorts of jobs in the RAF, often doing jobs alongside men which could equally well be done by men, and which require a high degree of calm and judgment.

At one station I visited in the United Kingdom I found (in a 'soft' brick-built office block of a particular squadron, but ready at a moment's notice to go into the adjoining hardened concrete block) a junior engineer officer attached to the squadron. She was aged twenty-five, very feminine, and wore elegant gold ball ear-studs. Her job included adjudicating on whether aircraft could fly with minor defects – a decision-making process which could be crucial to life in peacetime and to victory in wartime.

'One of my responsibilities is to ensure that a fault is not serious,' she told me. 'That might involve looking at an air frame for myself. They are old aircraft sometimes, and they often suffer from metal fatigue. Certain components are allowed "small" cracks. If they find a new one, it might be necessary for me to say that this is or is not within limits: if this part failed it would or would not present a threat to the safety of the aircraft, or anyone underneath it. It would be for me to say whether the aircraft could fly, or would have to wait for another component.'

For example? Well, a fairing might crack. This would not affect the strength of the main structure of the aircraft. If one way of holding it on to the aircraft snapped, it might be held on still by alternative means.

'They selected me in direct competition with men,' said the junior engineering officer, who had naval parents and an education at both comprehensive and boarding schools. 'I was trained with no special privileges, apart from the fact that, physically, I don't have to run two miles as fast as the men. It is a front-line job; front-line jobs are done by women in the RAF. There has never been any male chauvinism – in being an engineer, it is the same for non-commissioned women as well. They can work on duties on aircraft all round the world.'

This sounded a little too rosy. Was there any way the RAF could, in her view – after three and a half years in the service – improve its use of women?

This produced a disciplined avalanche. 'For a start, we don't get equal pay, because we are not combatant. But the men here in this front-line unit – apart from the aircrew who take their flying machines into a particular war zone – are in the same position as us. It is compulsory for us, bearing arms – a sub-machine gun and pistol. The difference between us and the men is that a man engineer in this job gets two and a half per cent more (the X-factor). I think this is unfair. If it was the other way round, there would be a massive outcry, but of course the RAF is exempt from the Equality Act.'

Anything else? Two things, she said swiftly. Until

recently, you couldn't raise a family *and* remain in the service. Obviously, in a lot of jobs it was impossible to continue in the service, but many servicemen had got children at boarding school, or capable of being looked after by other means. In these cases, experienced and trained women could be air traffic and administration officers, living on camp in quarters after re-entering the service.

There was, she said, a clause in the conditions of service saying that, if you didn't expect to live and work with your husband, and if you wouldn't use your child as an excuse for not going anywhere in the world, you might have a family and stay in the RAF. But the RAF didn't provide any means of looking after families. It was accepted that every woman in the United States Air Force could remain in service all their lives, but they got special benefits and help in looking after their children, which enabled them to take advantage of their entitlement.

And her second suggestion? 'The last male bastion! There are no female aircrew, apart from female air loadmasters – a small number of these are to do with flying. If you are an air loadmaster on a passenger route, it is equivalent to being an air hostess. You look after the passengers, feeding people and making sure they know the escape routes. On a freight route, you look after the cargo, making sure it is safely packed and balanced – acids, explosives, the lot. You make sure it is tied down securely; that it is appropriately packed so that it can be unpacked in the right order at the other end. It can affect efficiency and the safety of the aircraft. You may have to supervise parachute drops of freight. That is a responsible job, and women do it. But they cannot become pilots or navigators.'

The junior engineer officer said she had had to get a degree to get her job. At the beginning of 1986, there were only eighteen women among a total of 2,200 engineer officers in the RAF; the number had later gone up to thirty. But there was room for more women like herself, prepared to go, as she had done, to six different countries in one year – 'Flexibility is the name of the game.'

Another engineering officer, even younger at twenty-three, a quiet girl with very neat hair, had no complaints about her work in the RAF, but was a little conscious of social isolation – an admission that was jumped on with surprise (and perhaps hope) by some male officers who happened to be present in the mess when we had our conversation.

'If you want female sort of talk, you are restricted', she said. 'There are only four of us living on base. It is *very* isolated. It is not a great difficulty, you get used to it, but you feel lonely at times. My parents are in Kent and I can't drive, so I haven't been home since I went to college. I go to London, where I have a few friends (a journey little short of one hundred miles), or I go to see my boyfriend who is in the RAF. I go by train and it takes five hours.'

Then there was the senior aircraftwoman of nine years' RAF experience I found in the hardened logistics and engineering operations room of another base, in front of a computer screen. She told me her job was vetting demands for equipment and spares. The storeman of an aircraft servicing flight might call up and place a demand for parts of an aircraft – it could be anything from nuts, bolts and wire to a radio. Only priority demands would be dealt with instantly in her room. Anything routine would go through another channel, the main supply squadron. And it was up to her to judge, on laid-down criteria, whether each case was priority or not. Anything to do with an aircraft would tend to be given priority, because keeping the aircraft rolling was the chief priority. She said she would have to adjudicate on an average of about twenty demands a day; the work was responsible and testing.

One non-combatant activity in the RAF which might – at least to civilian observers – seem ripe for invasion has, in fact, resisted it completely: being a musician in any of the RAF's various bands. Unlike British symphony orchestras in civilian life, who have vastly enlarged their proportion of women players since the Second World War, the normally innovatory RAF has flatly refused to budge on its principle that being a musician is a men-only job.

Senior officers appear a trifle uncomfortable as they advance the reasons – as intelligent men, they are quite aware you may think they are rationalizing simple prejudice. When I broached the subject to one senior officer in the Music Services, an interview which had been congenial and easy became more akin to the extraction of wisdom teeth.

There were about three hundred men in the Music Services including officers and musicians and some civilian support, said the officer.

And any women? 'No, there are no women employed in the Music Services of the RAF.'

Why? 'The decision to disestablish the Women's Band of the RAF was made in 1971 – and that was a complete women's band, no men in it.'

Why? 'In line with current defence policy, women are only employed in bands as a total entity. There are no mixed bands, though the Army still maintains a women's band. There are no mixed bands in the British forces.'

Why? 'I would prefer someone else to give you that answer. There are problems. You can't use the emotional problems – i.e. the problems of mixing the sexes in accommodation and transport. You can't use this as an argument. You can't use the argument – because it is not valid – that women are not such good musicians; it is just not true. But there are physical problems which present themselves when, for instance, the Central Band goes to the Lord Mayor's Show, where they are required to march for three hours through the streets of London – that is physically a very demanding job.'

Too hard for *all* women? 'In some cases a women's band could go at their own pace, but unfortunately at a Lord Mayor's Show . . . No, strike that out, I'll start again. If the Central Band is employed at the Lord Mayor's Show and marches through the streets for three hours, they have to march at one hundred and twenty paces a minute, thirty inches to a step. There are even some *men*, if they are small, who find that difficult.'

Then couldn't women be excused certain jobs? 'It would be wrong to make a provision for a woman who normally plays the tuba to take it away from her for a particular engagement. The ceremonial role of British military bands doesn't really allow the idea of mixed sexes forming the band.'

But surely women could play as a matter of routine in concert performances and as a matter of routine *not* at ceremonial functions? 'I have to admit that I find the sight of mixed musicians in a *concert* role very attractive, because they are there just to use their musical skills as appropriate. It is recognized, of course, that flautists or clarinetists are auditioned for their musicianship in a concert role. It is only the ceremonial role, in a mixed orchestra, that causes problems.'

Did all countries find the problem insuperable? Hesitation, much crossing and uncrossing of legs. 'One recognizes that in America and on the Continent there are mixed bands. But it is something that defence policy will not accept in this country.'

Did he personally agree with that policy? 'I agree with that policy. The prospect of Guards changing guard at Buckingham Palace with a mixture of skirts and trousers of all shapes and sizes is horrifying – he says with a laugh.'

But the Americans didn't mind? 'I like the solution of the United States Air Force Band, which is something I would love to adopt. That is, within the establishment there is a concert band *and* a ceremonial band and the ceremonial band is a male contingent; the rest are judged by their ability. What I would like . . . what I would *love*, is the idea of women playing in the Royal Festival Hall because, simply, you have got the best person there. But when I put on my busby and boots and go out on parade on the streets of London, I prefer to have an all-male attachment. But that is purely a personal view. Can we have a change of subject?'

I obliged, afterwards reflecting that there was one other factor this particular senior officer could have mentioned but hadn't. Like all the ancillary services in the RAF, musicians have a separate wartime role. In the case of musicians, it is to be armed guards. This could marginally tell against employing women as musicians in a mixed Central Band of the Royal Air Force – at least until it is remembered that women are used as guards already, elsewhere in the British forces. I vividly remember the women guards I met at a Polaris submarine base when researching my book on the Royal Navy, *Ruling the Waves*, their hands hovering near their holstered pistols at the sight of a strange face: heaven help the man who tangled with them. It is an irony, but apparently true, that although a woman might not be as good as a man as a nightclub bouncer, when brute force and minimum injury to the baddies are crucial, they can be just as good as a man when the odds are higher, guns are in order, and avoiding damage to the baddies is not such a crucial factor.

Perhaps, then, that uncomfortable senior officer in the music services was saying no more nor less than the truth when he said it was the ceremonial aspect – that need to present an even, consistent line of bodies – that was crucial. All the same, the thought of a unisex concert orchestra or even chamber group seems inviting – especially as there is already a small orchestra, drawn at present almost entirely from the men of the Central Band and reinforced with string players, who can give chamber concerts.

But the inability to become a musician affects only a minority. Pay, which came up widely if not deeply as an issue (there were movements towards reform in the pipeline), is more central.

There are in fact two rather British subtleties, or bits of cosmetic surgery, that affect the careers of women in the WRAF. The first is that though women are technically commissioned into the RAF, not a separate body, on joining they immediately find themselves part of the WRAF. The second is that women, however they are described, in theory earn exactly the same as men in a similar job, but in *practice* do so only partly. The only exceptions to this are doctors, dentists and the legal branch who, whether male or female, are in the RAF, and the women earn *exactly* what men working alongside them get.

The explanation with which the rest of the women in RAF blue have to be content is the X-factor. This is a pay weighting for 'abnormal-type duties', as one senior WRAF officer put it. Men receive an X-factor of 10 per cent on top of their basic salary for a parcel of special factors, including working unsocial hours and facing combat. In the case of women, this parcel cannot include combat, because it is British government policy that women are not used in combat roles. Hence, at the time of writing, women receive only 7.5 per cent X-factor pay. The difference in the case of the highest-paid woman in light blue, the Director of the WRAF, an Air Commodore, is about £200 a year. In the case of junior officers and aircraftwomen it would not amount to £5 a week. But the differential is nevertheless regarded by some women as an affront to feminist principles.

The pay differential between men and women is perhaps one of the smaller problems faced by the 700 officers and 5,000 airwomen of the WRAF, but in the late 1980s moves were afoot to close the gap between what men receive under the X-factor and what their women colleagues get. In 1982, the percentage received by women jumped from 5 per cent to 7½ per cent, and by 1987 pressure was growing for at least another half a per cent or one per cent.

Why not 2½ per cent, giving women absolute equality with men? I put the question to the Director of the WRAF in her office in one of the furthest-flung buildings of the Ministry of Defence, in a rather anonymous road near High Holborn. She was a short, energetic lady with a commandingly large single string of pearls, a designer blouse with blue, grey, orange and yellow stripes, immaculate black court shoes, a warm handshake and a history of energetic squash and tennis which, she admitted reluctantly, was in the process of being transmuted into beginner's golf.

'I would hope for a further movement *towards* 100 per cent for women,' she said cagily. 'But I am doubtful about

100 per cent itself, because of the combat factor. One hundred per cent might not seem equitable to the men who have to go into the combat roles. At the moment there is very little aggro between men and women; we are superbly supported by our male colleagues.'

So the crucial question was: would women ever be likely to fly in aircraft, even as pilots?

The Director pointed out that women already flew in one role, that of air loadmaster, the person responsible for seeing the aircraft was trimmed properly, all the passengers and freight being located to achieve the proper balance.

I pointed out that during the Second World War, women had flown aircraft from the manufacturer to an airfield, or from one airfield to another, acting in effect as deliverymen. Why not allow women to do this today, especially in peacetime?

The reason, said the Director, was partly an economic one. The RAF recruited its pilots and navigators to fly fast jets and occupy the front-line combat seats. From those who did not meet the stringent criteria for this, it could recruit people to fly multi-engined aircraft, like the Hercules transport aircraft, and helicopters. Because of the rule that women could not occupy combat seats, if they were allowed to be trained expensively as pilots and navigators, they would have to be limited to the non-combatant type of aircraft and kept out of fast jets. That would not be cost-effective.

Did she imagine the situation would ever change? In a way it *had* changed, she said. Regulations for the University Air Squadrons, which trained aircrew, had been changed in the past couple of years. Some people who were joining the RAF in the ground branch could now fly in the squadrons at about eight leading universities. This meant that women had the chance to fly aircraft as well as men, which could greatly help them in many ground staff jobs, including fighter control.

The chances of equality for women air loadmasters had improved, too. Up till about six years ago, women air loadmasters were allowed only on to RAF aircraft which were standard passenger aircraft in the civilian sense, such as the VC10s. Now they could fly as air loadmasters on a much wider range of aircraft, such as the ubiquitous Hercules, which could carry both passengers and military freight.

'It's a gradual evolution, though,' said the Air Commodore. 'Nothing is a revolution in the service.'

She said she could remember only two jobs she had filled in her entire career which could not, within RAF regulations, be filled by a man. They were Director and previously Deputy Director of the WRAF: these had to be filled by women.

Her career progress was certainly interesting as an example of how a woman with no military background (father a chartered surveyor) could rise to a £32,500 or so job in the modern WRAF. She joined late, at the age of twenty-seven, after having a number of secretarial and administrative jobs on the continent of Europe. Two years older and it would have been too late for her, she thought. She joined as an officer cadet, doing three months' basic training and another three months' professional training before taking up her first job as ADC to the Commandant of the College of Air Warfare.

'A man could have done that job,' she said. 'I suppose I got it as my first job because I was twenty-seven – you could hardly have had a girl of eighteen or nineteen in it.'

She then trained as an Air Force accountant, going first to Brize Norton transport airfield and then to Libya as a junior accountant. Her next two postings, both in the United Kingdom, concerned training officer cadets. They were followed by accounting duties in Holland, a personnel job at an RAF station and a period as a student at the Staff College at Bracknell, the RAF's school for those professionals deemed to be on the fast track. She stopped there as a member of the staff. Next came a job at the Personnel Management Centre as a staff officer handling the progress of officers' careers.

From there she became officer commanding the Administrative Wing of a flying training station – 'seeing to everything, including education and physical education, except those things that fly or need spanners'.

Next came the deputy directorship of the WRAF, followed by the directorship – after which there was nothing further for her, except retirement.

Of course it might be dangerous in general to infer the chances the RAF gives to women from the career path of one bright woman. Was she the statutory successful woman, a totally untypical example? I put the question. She admitted that, of course, only one person could be Director at any one time, but thought her career had started quite ordinarily, with promotions open to any intelligent, outgoing woman who worked hard and played hard.

'A lot of it,' she said, 'is the luck of the draw.'

Was it also something to do with marriage? She herself was unmarried. Could she have risen to the top if she had faced the conflicting demands of a husband, a home and children?

I got the reply: 'My deputy here is married. As an officer I don't think it makes any difference whether you are married or single. But I must say my deputy is married to an RAF man. Most of us tend to marry RAF men. Very few of us marry Army or Navy men or civilians. It certainly makes it easier to marry within the RAF because although we don't guarantee co-located postings we do try wherever possible to do it. It makes good economic sense. That way we don't lose people.'

Statistics bear her out. A decade ago, 15 per cent of WRAF officers were married and 17 per cent of airwomen. Since then, the percentages have jumped to 30 per cent of officers and 36 per cent of airwomen. Women no longer regard marriage as an almost automatic reason why they should leave the service. There are two reasons, clearly visible in society, why they might be more willing to stay: the dearth of employment in civilian life and the fact that increased home ownership means that two salaries are definitely better than one when it comes to paying off the mortgage.

I asked the Air Commodore what sort of woman the WRAF was now looking for, as compared with the other forces. 'I don't want to give you the wrong impression or be pompous. We are basically all looking for the same type of person – who mixes well, is intelligent, is outgoing and who is genuinely seeking a career. We are all *not* looking for a recluse or a shy little soul who is only interested in a nine-to-five job. It is very competitive to get in. I don't want to set the RAF on a pedestal, but I do think in the end that women are better integrated within the RAF than they are within the other two forces.'

Yes, but she would say that, wouldn't she? 'No, there is nothing magic about it, but we are in effect totally integrated, not a corps like the Army or a structure like the Navy. The Navy goes to sea and the women are precluded from doing that because of the combat situation, which it is national policy that women do not take part in. The Army goes to war and, because of the combat exclusion, Army women aren't part of that. Whereas we do send our chaps to war – but a very few of them. The vast majority of the RAF is the tail rather than the teeth. Therefore it is easier of us to be totally integrated into that body.'

Was I right that, in this and other respects, the RAF seemed to be more 'civilian' in its atmosphere than the other two forces? The Air Commodore in charge of the WRAF said she sometimes thought she could have become a doctor if the RAF had not been as fulfilling as it had been – and I believed her. Her personal manner was certainly more comradely than lordly.

'Yes,' she said, 'we are more *producers*, *workers*, as RAF people. The Army is all about man-management. Just about the only people in the RAF who have a man-management role are the engineers. They can be in charge of dozens of people; hundreds. Most of the rest of us don't control very large numbers of people. And quite a high proportion of our officers now come up through the ranks.'

Women may be making progress through the ranks of the RAF, but the women who are the *wives* of RAF men tread a fine line between being independent of the RAF (which they like) and being left out on a limb (which they don't like quite so much).

When they are abroad, wives of RAF men (like women members of the RAF) tend to find that both the good and bad points of RAF life are heightened. It certainly applies in Germany, regarded as one of the best postings: especially, perhaps, to wives of men stationed at RAF Gütersloh, the only base east of the Rhine and the nearest to the Eastern bloc forces. Men inevitably have their minds more on the Harriers, Chinooks and Pumas than on their families, at least at times; inevitably the wives are thrown back on their own social groups for entertainment. And here the social line between officers and airmen, and officers' wives and airmen's wives, can be a complication.

A further complication is that quarters at Gütersloh are scattered and mostly off base, sometimes miles away. When I talked to a group of two officers' wives and two airmen's wives at Gütersloh, I was left in no doubt of the things most on the ladies' minds (they did *not* concern mere local complications); and I shall use fictitious names to avoid any embarrassment.

Mrs Jenkinson, wife of a Flying Officer, said she liked Gütersloh because, on her previous posting, there had been an RAF community of only about twenty people, whereas in Germany there were more people and more things to do.

That immediately set off Mrs Pringle, wife of an administrative clerk.

'Squadrons are very far away from the camp,' she said. 'We don't have a camp life as such. We don't go to the camp to have a social life. We go elsewhere. If you want to go to camp to have a drink, you have got to take a taxi. While most of the officers' quarters are near to camp, I am nearly thirteen miles away. The difference between officers and men is money. If it was not for the fact that I can work as well as my husband, we could not afford the things we have now. A car, for a start. That would be a further restriction, if we did not have that.'

This far, I could have thought that what I was about to hear was predictable: the officers' wives content with life, the less privileged airmen's wives not quite so happy. Not so. The most vocal officer's wife, Mrs Jenkinson, and the most vocal airman's wife, Mrs Pringle, were soon drawn into unanimity by my next question: how did the RAF as a whole cope with wives?

Mrs Jenkinson replied instantly. 'I think very poorly, actually. It is something you have got to accept. I think wives are treated as second-class citizens, you are never a person in your own right. For instance, the library. You aren't allowed to join in your own name. If anyone wants to get in touch with you about a book, they get in touch with your husband. So, in other words, you haven't got a secret or separate life from your husband. Does it rankle? Yes, it does.'

In came Mrs Pringle like an Exocet. 'If you want to buy something from the Naafi, you have to give your husband's details, even if you are paying cash. If you are given a receipt, it is not "What's your name?" but "Who's your husband? It's your husband's details I need!"'

Mrs White, wife of a Flight Lieutenant, had not spoken so far. Now she thundered: 'The wife of! The wife of!'

The other ladies all made cries of approval. It was the phrase, I was soon made to understand, that RAF wives hear most often, and the phrase of which they were most heartily sick, because it implied that they existed only in terms of their husbands.

It was a complaint similar to those I had heard from Army wives when researching *Soldiering On* and Navy wives when researching *Ruling the Waves*; and I had been very careful, when dealing with the RAF wives, not to ask leading questions on the point. But I was hardly surprised when it emerged spontaneously and almost instantly – quickly overtaking and overlaying all other grievances.

'I was even told I had got a job – in education – through

my husband,' said Mrs Jenkinson. 'My husband had to come and tell me – it was my husband who got the information. That is just an example of the way everything that concerns you is told to your husband, and he is expected to pass it on to you. It changes a woman's life-style, being married to an RAF man, more than it would change it if she married a man in any other occupation I could think of. To cater for that point, an awful lot of changes would have to be made in the RAF. The only way *I* can cope with it is to think it was my *choice* to marry him – he was in the RAF when we married. My husband makes the same comments as I do. His feelings about it are the same.'

'Things have happened exactly the same with me,' said Mrs White. 'Recently the medical people did exactly the same with me. A few of us had a moan and groan about it. It was pointed out that we didn't want to be "the wife of". Now I give my own name and tell them the address of the quarters so they know they have got the right person.'

This produced a chorus of agreement, in which it was pointed out that it was seldom that husbands, even in the RAF, got pregnant or had menstruation problems.

Mrs Jenkinson recalled another example. 'I came here after my husband. You need to pick up an identification card. The office rings to say it is ready. I was not informed myself, when I picked up the phone, that the card had arrived. I was informed, "The card of the dependant of Flight Lieutenant Jenkinson has come through." That rankled. Station routine orders only go out to husbands. They should make an attempt to get in touch with wives.'

The 'wife of' syndrome was virtually the *only* point on which all these four wives agreed. Mrs Jenkinson said her husband was on a fifteen-year engagement, with twelve still to go. Would she, because of her feelings, encourage him to leave sooner or not renew the contract? No, she said, she wouldn't go that far – not at this stage, 'unless I came across something that would *really* rankle with me – these are trivial things, I suppose'.

The wives praised the welfare back-up that was there to

cope with the problems of young wives who missed their mums, or were stuck at the top of a block of flats miles from the base. But Mrs Pringle, wife of the airman, showed all too clearly the problems an officer's wife can face if she tries to help airmen's wives. 'Airmen's wives don't want officers' wives telling them what to do,' she said. 'If an officer's wife came to me, I would feel she was interfering in my life, and I would tell her where to go. And, as for other airmen's wives, you can't have a friendship with them, because in a few year's time you or they will be leaving.'

Mrs Brown, wife of a Sergeant, pointed out that a senior NCO's wife could be involved in helping young wives of airmen.

Mrs Pringle wouldn't have that, either. 'I don't like someone I don't know very well saying, "Have you got a - problem?" It isn't private then, because that person will probably tell everybody. If I were to say I was having problems with my marriage, my husband would find out and half his section would know. I find it nosey.'

Mrs White gently suggested that only some people were nosey. others genuinely wanted to help. Mrs Pringle wouldn't have it. 'More than likely I would take advice from men of my husband's rank. Officers and airmen are so different. Officers are not approachable. If I had a problem in work, I could not go to one of them and say anything. Where I am a clerk, there are four officers and one Flight Sergeant. I would not go to an officer, because I would not think I could speak to him.'

It was by now striking me as distinctly possible that Mrs Pringle had personal hang-ups about social position in general. It appeared that her father had been in the RAF for over thirty years, and was now a Warrant Officer. Her sister was also in the RAF – as an officer, a Flight Lieutenant. 'She is upper-class and I am middle-class,' said Mrs Pringle. 'I don't think I am excessively class-conscious.'

I said nothing.

Mrs Jenkinson supported her – up to a point. 'I would think that attitude is average in general,' she said. 'But I would not say it was the *right* attitude.'

Mrs White pointed out that there was inverted snobbery as well as ordinary snobbery. Her husband had come up through the ranks to be an officer, and when he had got his third stripe her next-door neighbour wouldn't speak to her, because *her* husband was still a Corporal: 'She simply snubbed me.'

Mrs Jenkinson said she had found something similar: she got more respect as the daughter of a senior NCO than she did as the wife of an officer.

I broached another touchy subject: married quarters. 'Only if you've got half an hour,' laughed Mrs White. I told her I *had* got half an hour, but, while she was thinking, Mrs Jenkinson regretted that she had a poky little kitchen with two cookers, one with the door hanging off. I asked what the decoration was like. 'There *is* no decoration. Green emulsion everywhere.'

Mrs Brown said she was given a choice of colours: for the living rooms, white, ivory or magnolia; and for the bedrooms, green, pink or blue.

Mrs White had now worked out what it was that *really* upset her: the 'march-outs'. In the RAF, as in the other armed services, there is a system under which a quarter is officially inspected before the occupiers are marched out and the new ones marched in.

'We, as women, are not allowed to march out. So, if your husband is already away at his next base, you have to get a friend, who is senior, to do it for you. Those making the inspection inspect all the house – I have even had one families officer turn up with white gloves on. He went to the lampshades to see if there were any cracks, and the teapot to make sure there were no chips, and the cooker . . . I have been charged for four hours' work to a cooker if there was any grease, but I am sure it was never actually cleaned for the next person. A wife cannot accept a new flat, either – the man has to be sent back to accept it.'

The wife of a Department of the Environment worker on the Gütersloh base, who had been sitting in on the conversation, chipped in with her own anecdote. 'Even the linoleum has to be the same colour as the linoleum under the carpet, despite the fact that the daylight will make it fade. Even without children, it is impossible to make it look as if the place hasn't been lived in.'

A certain guile is required among wives to deal with such points. Mrs Brown recalled having a coal fire in one quarter, with the result that coal dirtied the carpet near the fire. 'We couldn't get it quite clean; but when it was wet, it didn't show. So we made sure it was wet when we did the march-out.'

But it was not such material matters that dominated the conversation for long: soon we were back, without any prompting, to the 'wife of' and 'officers and their ladies, airmen and their wives' syndromes, with Mrs White pointing out that if wives were happy, the men were happy at their work, and vice versa. 'I was even introduced the other week as "Mrs Flight Lieutenant White". I immediately got up and said, "*I* am not in the RAF." I was told, "You know what I mean." I said, "I know what *I* mean."'

I had to drag the conversation away from this issue, by asking for other observations about the life of wives. Mrs Pringle was irritated because she had had to get married not in *any* church, but in the station church, an attic with dormer windows. Mrs Jenkinson said that RAF wives, after a divorce, could be left homeless, moneyless and helpless; and that, generally, there was in Germany a temptation to spend money on an expensive car because you could buy it at a cheaper rate. Mrs White regretted that the West German mark, once twelve to the pound, was now down to between three and four.

Very little concern was expressed by any of the wives about the precise role of the RAF in Germany, a role that (though not through the Harriers at Gütersloh) could include nuclear arms.

Mrs White, the gentlest-voiced of all the group, at last addressed herself to this point. 'To be honest, if I really thought about it, I would get frightened of their role here. I got stuck on the camp once. I was working here, and the alert went out. Within five minutes that road out there was crawling with men in their nuclear–biological–chemical coats. It put the fear of God into me. By the time I was down that road I was all in a do-da.'

The wives know the drill if there is a real emergency: they would be taken to a port and sent back to the United Kingdom. They would be taken by the RAF in groups, because if they went individually they would be classed as refugees. Mrs Pringle, ever the maverick, said that if it came to the point, she didn't know whether she would want to go or stay with her husband. The other wives felt that the pressures would be so enormous that they would do what was expected of them.

When I talked to officers' wives in RAF Brüggen in West Germany, a hundred miles or so back from Gütersloh, but a station with a possible nuclear role, as its Tornados are capable of carrying nuclear weapons, the conversation did not once shift to the physical dangers of the men's task.

Mrs Morrison, (once again, the names will be fictitious), wife of a Squadron Commander, said the only time she took part in the life of the RAF was when she was in Germany. But this, she said, was usually a matter of arranging coffee mornings and suchlike activities when the men were working or on attachments. Any social or other guidance of airmen or their wives usually came from the Warrant Officer or his wife, rather than an officer's wife.

Mrs Morrison, when I asked her about how she regarded the role of a Squadron Commander's wife today, said she made a point of visiting all new wives when their husbands joined the squadron. 'I visit officers' wives, other ranks visit each other,' she said. She thought it 'nice to belong to a squadron, which can be a replacement of the family'. She admitted that 'you take on the RAF when you have an RAF husband', but protested she did not mind being 'the wife of'. It was just the way the system worked, she said.

Yes, I thought – but you *are* the Squadron Commander's wife, so perhaps it is easier for you to accept the RAF protocol than for wives of men lower down the promotion scale. I asked her whether, on the whole, she would like her forty-year-old husband to stay or leave the RAF at the earliest opportunity.

'It fluctuates with every tour,' she confessed. 'If it is dodgy, he thinks about going; if it is good, he wants to stay. *I* would like him to stay. My only feeling is that, as your family gets older, you want to be back with them. Our children are now going through O-levels and you feel you want to be near the family. But I enjoy the life: I suppose I am a camp follower by temperament.'

Then I asked the same question of the wife of a Flight Lieutenant in the same squadron, Mrs Adams, a woman whose striped brown shirt, blonde hair and pancake make-up make her look even more trendy than the tall and elegantly dressed Squadron Commander's wife.

I was not entirely surprised when she was more guarded in her views. 'He has got another eight years to go. At the moment, I think he should think seriously about having something else behind him to do if he should decide to leave. I want to study law and he wants to open a restaurant. I feel it is time he starts to think about whatever else he wants to do, and which, if things aren't going well, he can get to grips with. It is no good leaving it until then and thinking, "What am I going to do?" At the moment I am enjoying it very much. But we would like to look and plan ahead.'

Mrs Adams, it struck me, was like a lot of intelligent young wives of RAF men: she was mentally reserving her own position and was trying – not altogether successfully – to put her *own* feelings into the background.

After the Squadron Commander's wife had remarked, 'Germany is a hard tour, because there is a lot of entertaining and social things we have to do; it would be nice to lead my own life for a while; I am looking forward to a rest', Mrs Adams gave a friendly smile.

'I get along with Mrs Morrison fine,' she said. 'We are on first-name terms, whereas with some, it would be Mrs So-and-So all the time. It is the individual concerned, not the rank. I have come across a lady who was very overpowering and absolutely dreadful. Every time you opened the door, she was there, putting pressure on you to do things. Whereas with Alice, she doesn't apply any pressure. You *want* to go over and get involved. That would be my technique. I don't think I would come on strong – it is offputting.'

I asked Mrs Adams which was the more typical – the discreet Mrs Morrison or the overpowering lady?

'I have had two nice ones and one overbearing one,' said Mrs Adams.

Perhaps that is a proportion that could not be bettered in any civilian occupation.

Among officers' wives, the desire *not* to seem overpowering can often strike a civilian as a little sad. At the Rheindahlen joint headquarters, which the RAF shares with the British Army, I met one wife who timidly confessed she did voluntary work for a newly created senior citizens' club at the headquarters. This, she said, was primarily for airmen's relatives, fathers or mothers who had nowhere else to go but their serving sons' homes in the RAF.

'I just organize get-togethers in the church centre,' she murmured so self-effacingly I could hardly hear her. When she had first answered the door, I had diagnosed her as a hearty lady, largely because of her large, flamboyant pearls. But immediately the subject of welfare work and officers' wives was raised, she became a shrinking violet.

'It is difficult,' she whispered, 'because every generation, except this one, has been able to cater for its old people. It is more difficult for this one, and perhaps even more difficult for those who are in the forces. There are about six of them, aged from sixty-nine to seventy-six.'

I asked her what she did for them. Immediately the first person singular disappeared and 'we' took its place: 'We get them together and ask them what they would like to do. We have things planned for theatre visits within the camp, and visiting outlying areas of the camp. Within reason, we will take them out to see the country.'

What sort of feedback did she get about their problems?

'Oh dear,' she said, 'I wish we hadn't raised this subject! Perhaps it was the wrong thing to do.'

It was plain she was terrified of being thought patronizing or condescending. 'It is not so much their problems they come to discuss – because they have their dignity. They come for a chat about what it was like during the war, with rationing and so on. It is very interesting to hear them talk. They are all in the same boat; there is no day-centre at which they can meet. We suggested that this should happen, but I don't think any officer's wife is pressurized into doing anything she doesn't want to. We may feel it is our duty towards our husbands.'

I asked her if she had ever become involved in welfare activities for airmen's wives. The answer was oblique but perhaps revealing. 'They keep their lives very quiet, really – increasingly so.'

Mrs Blake was in a position to make comparisons, having been in the RAF herself and enjoyed it. Her husband made the transition from the junior ranks to a commission, at the age of thirty-four. She said she had found that very difficult, possibly due to under-confidence on her own part. She still felt there was a barrier between airmen and officers and, to a certain extent, between their wives. There were different rules in the messes. In the officers' mess you could not, as an officer's wife, wear trousers in the evening.

But, said Mrs Blake, there were more human difficulties. The problem of old friends who were NCOs or NCOs' wives could be difficult. 'I still *write* to people I have known for a long time. But if they are on the airmen's side, and they want to come to visit, *it isn't done*. You could probably get away with it if they were from another base and no one knew who they were.'

I could not visualize any self-respecting person making a visit on such a basis. Mrs Blake agreed. 'The barrier is often from the other side. My husband and I met up with a couple we had known when we were in Cyprus. Both might have come down here a lot. But it was difficult for her husband to come and socialize with us, because he was still an airman. It was very difficult. It didn't work out. We stayed very good friends, but the husbands were in the background, because one was an officer and the other wasn't. There is a quiet rule against it. It doesn't work very well when husbands are working close together – socializing would not help. Personally, I believe that people are people are people. But I can understand that, from a working point of view, there has to be a separation.'

I asked Mrs Blake whether one mess, for officers, NCOs *and* airmen, would help. Mrs Blake thought not. To the men, the officers had to 'appear strong and good' at all times. If there were only one mess, it would not be easy for them to let their hair down, and it would not be easy for everyone else to relax in their company.

When lives may depend on the instant following of orders, there may be a case for the caste system between officers and others. But there is one section of the RAF where such pressures are not so intense, and where in fact the *only* way a man can become an officer is through the ranks, meaning that all officers know what life is like on the other side of the fence. This, oddly in view of its total rejection of women in its conservative ranks, is the world of the RAF musicians.

The wife of a Squadron Leader Bandmaster I met in Rheindahlen joint headquarters, a merry north country lady with a comfortable home full of snake plants, Bavarian figurines, Lowry reproductions and pictures of her grandchild, said her husband's move to officer status had not been difficult, and that she had not had to drop any old friends.

'We are in a different position to some others in the RAF,' she said. 'The people you are in charge of are old comrades. I know their wives and don't drop anybody. A long time ago, it was frowned upon, but standards have relaxed. You no longer have to wear gloves in the mess, and officers' wives don't have batwomen to do the housework.'

Would it be better if *everyone* in the RAF had to come up through the ranks? Mrs Blake didn't think it feasible. And neither did Squadron Leader Blake: 'You haven't got the time. By the time he got to being pilot, his reactions would be as slow as mine. But it would be a good idea if senior officers went and found out what the rest of the RAF did.'

Back came Mrs Blake: 'You get slightly preferential treatment as an officer's wife. You give your address, they know your husband's rank, and they come around straight away to do any house repairs. Officers and airmen get the same-coloured sofas, but the officers get three-seaters and the airmen only two-seaters. On the other hand, we are *all* treated as "the wife of". You always have to give the last three digits of your husband's service number to explain who you are, and of course you can't remember. *But if you let that bother you, lots of things would bother you.*'

Where women and the RAF are concerned, that appears to be a widespread maxim. It also appears to be a maxim equally widespread in the RAF as a whole, both in the front line and behind the lines.

PART II

BEHIND THE LINES

7

THE BUNKER AND BEYOND

'We can run an aircraft up north to attack a shape on an electronic screen close to Iceland; and then, four hours after that, the same aeroplane can be directed down hundreds of miles south, and be attacking a target off Gibraltar. You have this enormous flexibility, which is why Strike Command must be centralized – you cannot have an Air Force in penny numbers any more. It is *not* a bureaucratic reason.'

– Senior Air Staff Officer (an Air Vice Marshal) at Headquarters Strike Command near High Wycombe, Buckinghamshire.

'When Rommel was going through France in the Second World War, he took time out to go home for the weekend. It ain't going to be like that next time. Our computer systems are being updated, and some crisis management decisions will be made by computer, because there is no way the human hand can keep up the pace with a fast flying jet bomber.'

– Senior officer, Headquarters Strike Command.

The neat little Chilterns villages in Buckinghamshire around High Wycombe are immediately at odds with what is often called the 'bunker' mentality. They have fresh air, neat hedgerows, newly gilded church clocks, well-tended cricket fields. But the bunker itself, with its nuclear and other potentialities, is real enough; as real as, just up the road, the Red Lion pub, the National Trust forests that nestle beside the winding lanes, the rolling village green ascending to the door of the squat grey-stone church, the rows of combed-and-brushed redbrick or mellow stone cottages.

Such is the picturesque English view as you approach the Royal Air Force Strike Command Headquarters – which guards United Kingdom air space, watches what the Warsaw Pact aircraft are up to over an area of 1,100 by 400 miles and has an electronic say-so over eight hundred RAF aircraft, ninety stations and two hundred units.

At first – and certainly from the air – the headquarters itself might appear to be just another English village. That is precisely how it was, from the first, *meant* to look. It was built at the start of the Second World War as the headquarters of Bomber Command; it was put on a spot where the variations in the colour of the soil in the surrounding fields are confusing from the air; and it was architecturally styled so that the brick building in which Air Marshal 'Bomber' Harris had his office could be taken for a community centre, the block of offices could be a school, and yet another block could be a chapel.

The Air Marshal's office was on the ground floor at a corner of a building; underneath it ran a passageway direct into the bunker from which the bombing of the Axis powers was master-minded. The bunker, almost entirely underground, then had walls covered with maps on which the position of the enemy could be plotted. Its main sounds would have been the clicking of pointers on the maps, the ringing of telephone bells and the shouting of instructions.

The bunker, I found, was still there. Only now it was filled with hundreds of millions of pounds' worth of electronic equipment: three neat levels of men and machinery within the space of a small provincial dance-hall. Most of the people were working in aptly named 'cells', in which specialist groups process information from all over the world and, in an emergency, feed it into what is *informally* known as the Commander in Chief's Cabinet Room and formally as the Battle Management Group Room.

In this room, the size of a tolerable domestic living room, the RAF's largest command makes decisions which, in frequent exercises, are demanding and which, in a real emergency, would affect what happened to Warsaw Pact aircraft which strayed from the accepted norms of behaviour.

This heart of the bunker is as reassuring in its apparent banality as a conference room at a computer firm or chemical company. The Commander in Chief's electronic screen sat to one side of his raised desk – actually an airman's dining-room table no longer required in a service being weaned away (as with the other two fighting services) from life in married quarters and towards home ownership. Arranged in a square, in front of his table, were another seven desks, each with electronic screens and keyboards – each one, in an emergency, to be occupied by the top specialists of the Commander in Chief's 'Cabinet'. About six feet away from where the Commander in Chief's right hand would be were wall-mounted telephone kiosks that looked utterly commonplace. They were in fact the lines to No. 10 Downing Street and the Cabinet Office.

These conditions were due to be modernized, in a new bunker about one hundred yards away from the old one; the job was due for completion before the end of the 1980s. The new bunker is four times the size, as well as being proof against chemical, nuclear and biological attack.

I sat in the Commander in Chief's chair in the Battle Management Group Room in the slender hope that this physical act might allow a civilian to feel just a tiny fraction of the sort of tension and concentration Strike Command's Commander in Chief might experience as he had to decide whether to order defensive action by one squadron against Warsaw Pact aircraft approaching over Norway and then take further action over Gibraltar, hundreds of miles south, before returning to base.

'It used to be said,' one senior officer in the Battle Management Group Room told me, 'that an army moved at five knots, and a navy at fifteen knots. The RAF now moves at twice the speed of sound; and so some operational decisions have got to be taken by computer, because the human brain and hand cannot move fast enough.'

But the basic decisions are still taken in the Commander in Chief's chair, which confers upon him a new sort of loneliness in command. Before Wellington, commanders tended to suffer the agonies of ignorance; of simply not knowing what was going on just out of sight. The loneliness of the top man in the Strike Command bunker is the loneliness of having *too much* available information, not too little of it. His specialists, in their little cells, are filters of information as well as providers of it; without them the calmest Commander in Chief's mental health might begin to deteriorate to the point where the 'dual-key' principle followed in the bunker became more than a formality.

In fact, the Commander in Chief can simplify matters by deciding what material, from what particular source, should be thrown at any one time on to a six-by-six-foot wall-screen, where it is visible to him and to all the other specialists in the room. Thus, at any one time, there are several electronic dialogues going on at once: between the specialist 'cells' littered around the bunker and their sources in various parts of the world (171 worldwide including some in the United States Air Force and the US Navy specialist cells, which include those of the other armed forces) and the top specialists' screens in the Battle Management Group Room; between any of those screens and the Commander in Chief; and between any material selected by the Commander in Chief for presentation for discussion to everyone in the room via the huge wall-screen. It is not an environment for the slow-witted.

Invariably the Duty Controller (who, in the wound-down state, is the Commander in Chief's direct representative) has to be inside one of the small glass-fronted cells at the bottom 'working' level of the bunker with his 'dual-key' man, a senior NCO. They could respond to a sudden threat well before an enemy missile would reach a dangerous proximity to Britain or any other target. They work all the time with a basic staff of about ten, which would be rapidly expanded in an emergency.

All this enhanced staff would have access to the vast electronic information wall which long ago replaced the maps, aeroplane stickers and long pointers. The wall takes up the whole of the three levels on one of the longer sides of the room; is (mercifully) totally incomprehensible to a civilian in its wound-down state but, in exercises, can get the adrenalin of the experts flowing in a passing imitation of the real thing. At such times, possibly for three days and two nights, the occupants of the bunker never leave it and (at least before the dormitories of the new bunker come into use) snatch sleep on any available floor, praying that the prescribed number of bars of soap and toilet rolls have been delivered.

'People can flap about toilet rolls just when you think you've got everything important under control,' said one of the staff responsible for supplies. 'People will soon tell you that toilet rolls *are* important.'

In the new bunker, I was assured, things would be smoother: even the delivery of vital toilet supplies. But even when it was no longer the nerve centre of Strike Command, it was unlikely that 'Bomber' Harris's bunker would be turned into a museum – as for example Winston Churchill's little office below Whitehall has become an historical shrine, pens, cigar stubs and all.

'I think we would immediately need the old bunker for over-fill,' said a senior officer. 'I think you can take it that it would never be a museum, because it has been changed about so much since 1942 that most of the historical interest has been swept away.'

In the bunker, a visiting civilian with an anti-military turn of mind (not that such a visitor would be likely to receive an invitation) might see a scenario for a remake of Stanley Kubrick's classic nuclear war film, *Dr Strangelove, Or How I Learned to Love the Bomb.* The people *actually* in the bunker, in exercises or in emergencies, warn themselves conscientiously against both the bunker mentality and – even worse – what they call the scenario mentality.

It was outside the bunker itself, at the comfortable desk

of the Senior Air Staff Officer, that the Air Vice Marshal dealt with the point. He was a large man who looked even larger than he was, with one of those strong cleft-chinned faces which can progress from sleepy relaxation to searchlight attention in about a tenth of a second. When I suggested that *this* environment, with side tables bearing cups of English afternoon tea, might seem so ordinary that a nine-to-five attitude could creep in at the very time when something nasty was happening near Iceland, and coming in our direction, his answer was simply to flick the switch of yet another type of 'television' screen on his desk top, positioned so that he could reach the controls without having, even, to lean sideways. On to the screen came a representation of the top of Scotland, with the Warsaw Pact aircraft to the top right in different colourings from the rest of the information. These distinctively coloured blobs are a fact of life all the time at Strike Command: *no* one is further away from them than a flick of a switch.

In a way, the constant indication of the *real* position of the Warsaw Pact air forces is itself a warning against the scenario mentality.

'You must remember,' said the Air Vice Marshal, 'that we operate a system that will work on whatever scenario pops up. That is why we tend to exercise. It is vital to watch and know what is happening out there, and to react properly and promptly. I can't remember that, at any time, the RAF has failed a reaction test. We take rapid reaction very seriously.'

But sitting in front of a screen on a non-exercise day, biting one's nails and wondering what an enemy might just possibly do at some time in the future, could do as much harm as good. 'You need a full spectrum of military force, because otherwise you can't respond properly. You want to be in a situation where you can hold an attack without escalating it. I don't think you will find a bunch of Dr Strangeloves who are behaving in a way likely to get their families worried. Not in the RAF. I personally don't believe in getting too scenario-conscious, because there are just too many moves a possible enemy could possibly make. I doubt, for instance, whether you are ever going to be able to second-guess the Third World War; and if you *could*, then the other side have proved themselves to be remarkably stupid, because the whole thing is to do something the other side *didn't* expect. You can't sit and worry about scenarios. Your systems must be adaptable, flexible, give command information from which they can work out the situation and draw up options, select the best one and operate the forces on the basis of that.'

It was a calming conversation to have in a smallish group of buildings from which the largest command of the Royal Air Force is centrally controlled and in which the safety of United Kingdom air space, and the United Kingdom itself, is vested.

It is ironic that the organization which is responsible for the safety of RAF resources all over the world is hardly in sight at all at Strike Command headquarters: it is not the RAF Regiment but the Royal Air Force Police at the gate and guardhouse who sort out whether you are a writer entitled to a pass, a concerned lady from the National Trust (also entitled to a pass), a harmless vagrant or a member of the Russian Special Forces travelling incognito.

The RAF Regiment, which guards all airfields and installations, would be there soon enough if an enemy special forces attack began in nearby woods. Otherwise its presence at High Wycombe is most conspicuous in the bunker, and at the highest levels of authority. Years ago, the RAF Regiment tended to be firstly told how the RAF proposed to build an installation and secondly told: 'Defend it.' Nowadays there is more consultation with the top echelon of the RAF Regiment on how stations and installations are designed, and how they are fireproofed.

This expanding influence is perhaps reflected to some extent in the chair available in the Battle Management Group Room to the Group Captain who is the Command RAF Regiment Officer. His job, in an emergency, is to tell the Commanding Officer what airstrips are still open, what airstrips have been taken out by enemy action, how long it will take to get them back into use and whether the aircraft returning from a mission should go to their own base or to X airfield instead. It is an electronically based catalogue of *cans* and *can'ts*, without which all the other top calculations going on in the bunker would be about as much use as the non-existent army divisions dreamed up by Hitler in the closing weeks of the Second World War.

The Group Captain of the RAF Regiment whom I met at Strike Command was a soldierly looking man with gold-rimmed glasses, lank fair moustache and a long history of training RAF Regiment men at Catterick headquarters. Now, he said, his horizons had expanded to include not only training but control of light armoured squadrons here and abroad, and the defence of all installations.

'I always say that I am responsible for seeing that flying *goes on* flying,' said the Group Captain.

He was at pains to show me how much he was attuned to the most basic practical considerations as well as to the larger ones.

He held up a green webbing belt from which were suspended the various items of kit that were ready at a moment's notice to go with him into the bunker. Tin hat (with camouflage), two ammunition pouches, a bayonet, a water bottle, some mess tins, bars of chocolate, a biological–chemical–nuclear suit wrapped in a poncho (which can be used as protection against rain or made into the roof of a low-ceilinged tent in the field) and a respirator.

The doughty warrior, who had joined up as a National Serviceman and was shortly to retire, allowed himself this unsoldierly thought: 'I rather hope we never have to wear the respirators too long.'

His reasonable point is being taken care of in the improved built-in protection of the new bunker. Also, it seems to be generally acknowledged that the next air war, if it comes, is likely to be short, with NATO air moves made quickly and powerfully enough to spell out almost immediately to an enemy: 'Stay precisely where you are, or . . .'

The Senior Air Staff Officer pointed out that all sides understood 'the very high lethality of modern air combat – it is getting bloody dangerous, because you are getting some very high-explosive, very agile combat missiles, and there are planes with high rate of fire guns, and very much more destructive guns. A hit is far more likely to give you a kill than in the past.'

Men flying aircraft and the men directing them from individual stations and centrally from High Wycombe – and men on the other side, too – must all be aware that aerial poker for *low* stakes disappeared when flying machines could no longer hope to creep home, 75 per cent of engine power gone, on a wing and a prayer.

It is a tense reality. But perhaps because it is so familiar and accepted at conscious and unconscious levels, the atmosphere at Strike Command – even in the bunker – is less blackly claustrophobic than it might otherwise be. It inevitably remains more tense than the atmosphere of the other Royal Air Force Command, which *backs up* the sharp fighting edge.

8

THE RAF PLC?

'If handing out some functions of the Royal Air Force to private contractors goes too far, yes, it will affect our capacity for war. The question is, has it gone too far already? My view is that it has. What concerns me as a serviceman is that it has gone too far for morale.'

– Air Commodore, Support Command Headquarters, RAF Brampton.

'The Treasury will pay people to leave the RAF, but won't allow us to pay to keep people in. We have had some people leave, taking their gratuity, and then coming back in again. I think a lot of things the Treasury do cost the Queen a fortune.'

– Senior officer, Support Command.

There is an inconspicuous eighteen-inch-square grey metal box with a combination lock on the wall of the Commander in Chief's office in Support Command Headquarters at RAF Brampton in the flatlands of Cambridgeshire. An ugly black tube leading from it disappears into the wall. It jars somewhat with the beige and saffron upholstery of the chairs grouped around the occasional table and the tinkling tea-cups.

It contains the secure telephone on which the Commander in Chief of Support Command, an Air Marshal and a knight, can talk unintercepted with 10 Downing Street, any British Embassy in the world, the White House, the Pentagon and even points East. It is six feet away from his desk and not much further from the occasional table and comfortable chairs around which guests are received.

There could be no more striking or effective reminder that, though the fast jets of Strike Command and RAF Germany may provide the teeth in time of emergency, they will in the end get precisely nowhere without the instant back-up of that majority of people in the Royal Air Force who make up Support Command.

But in the post-imperial last years of the twentieth century, Support Command Headquarters at RAF Brampton dances to the tune of economics rather than the immediate prospect of a Third World War. Under a decade ago, 15,000 civilians worked in the command, assisting nearly 50,000 RAF personnel in running what amounts to the housekeeping department of the RAF. The figure is now 11,500 and it is still falling, putting more work on to the shoulders of service personnel, who have increased in number by 2,000 in the same ten-year period.

Creeping 'contractorization' is taking over – the system of handing over specific support activities of the RAF such as catering and engineering to private firms. It is done on the basis that this costs substantially less – at least in purely financial terms – though many at Brampton fear that it could cost rather more in terms of reduced fighting efficiency in time of war when civilian labour would be more unpredictable than service labour.

Whatever its rights and wrongs – and the subject is a dominant topic of conversation in the mess bars of all ranks – the system is certainly changing the tone of life at headquarters, and at the 212 units around the world which it commands.

Sitting in the officers' mess at Brampton, one could conclude that the RAF was a baronial service with no money worries. The mess is located in Brampton Park, a fine country house dating back to the twelfth century. It was used as an institution for the cure of stammerers around the turn of the nineteenth century, until the place burned for three days in 1907. Later, in the First World War, it was used to house German prisoners and later still, in the Second World War, it became an American headquarters. After the war, what was then Technical Training Command took over. With the amalgamation of the Training and Support branches into one branch, it became the officers' mess of the united body, Support Command, which now gives the back-up to Strike Command and RAF Germany, dealing with administration, supply, training and numerous other housekeeping matters which concern the RAF as much as they would concern any private company. There are those at Brampton who sometimes feel that they could almost be referred to as the RAF Plc.

Among the handsome silver in the equally handsome hallway – recently restored to its former glory with clever reproduction staircases – there is at least one example of a lofty humour that perhaps belonged to a rather different RAF. It is a highly polished brass fire extinguisher, with a plaque bearing the words: 'I am the personal emblem of the President of the Mess Committee of Brampton Park Officers Mess. In the event of fire, abandon me and resign.'

Harmless facetiousness has rather gone out of fashion at Brampton, and throughout Support Command as it squares up to its main present-day preoccupations: shortage of money and government determination to farm out some tasks to private contractors who may or may not be used to life-or-death situations.

It is true that the atmosphere of Brampton does not have quite the tingling immediacy of life around the bunker of High Wycombe Strike Command Headquarters. The civilian worker administration office, beside the main gates, is a black-beamed thatched cottage, the streets within the headquarters are named after local bigwigs around Huntingdon rather than after fighting men, and some offices are located in ordinary domestic quarters accommodation.

But the problem of housekeeping can sometimes be quite as demanding as the problems of being prepared instantly to fight an enemy; and there are those at Brampton who fear that the great contractorization scheme, though excellent for saving money, may, if it goes too far, cause the *feeling* of being part of a disciplined and dedicated service to be diluted to an unfortunate and possibly even dangerous extent.

At the time I went to RAF Brampton, at least two supply depots had been handed over to private contractors. So had many of the support functions at one of the basic flying schools. On one station at least, both motor transport and supply functions had been put into civilian hands. Cranwell, the officers' training college, was due to have its engineering wing contractorized the following year.

From one point of view, the sense behind this was obvious. The RAF itself was being stretched in many ways: it was giving increased support to the United States Air Force at Molesworth and Greenham Common, and was taking over some facilities previously provided to the USAF by the British Army and the Royal Navy.

But was that the end of the story? I asked an Air Commodore at Brampton, a tall and solidly built officer with 5,000 hours' experience of piloting fighters, whether there were any snags.

His reply was eminently fair. 'Our experience with contractorization is that it provides good value for money, but there are unwelcome side-effects for the service as well, and there are limits to what we can do. In the basic training school which has been contractorized one is left with the Station Commander and flying instructors, plus the catering staff, in light blue, with the rest of it civilianized. They do the job extremely well, the civilians. But go and ask the airmen on that station what they think, and you get a different story. The service way of life has changed completely. Bear in mind the function of that station – to train future pilots – and you see the sort of start they have in their service career is not one we would wish. They are not being injected into a service environment. And that has happened at other places, too. You find that an airman can go to recruit training at Swinderby where he will see many civilians; from there he may go on to a station which is predominantly civilian, and be instructed by a civilian. He could spend the first five years on stations which are mainly civilianized – and that does not give him the best picture of service life.'

The Air Commodore recalled one station he had been on, where there was a flying wing with two or three squadrons, living in barrack blocks and enjoying service life esprit de corps. During the day the main objective was to teach people to fly, but between work and sleep there was ample opportunity for activities that were part and parcel of service life, like sport and doing guard duties.

'When the Dirty Dozen get rough and all our stations are on the highest alert state, where we have to control access to the buildings, locking every door except one and doing patrols and searches, we would then have to use what airmen had remained despite the contractorization,' said the Air Commodore. 'On some stations we are having to go back a bit on contractorization to civilians. I think we have convinced the financiers that you can't have a secure station where everyone goes home at five o'clock.'

I asked the Air Commodore if the policy of turning over large areas to civilians was based on the same unconscious premise that once had the Royal Navy using plastic-filled mattresses which would give off toxic fumes in a fire, merely because they were cheaper than horsehair ones: the dangerously unconscious premise that the service was a peacetime one?

The Air Commodore ducked that one: 'It is a way of making things cheaper in areas where we think, or they think, we can spare servicemen – because a serviceman is more expensive than a contracted civilian. That is the theory. However, we are not so sure, not in all cases. I will give you an extreme example of where they are wrong. At Molesworth, which is being guarded by civilian Ministry of Defence police, I think it would be cheaper to police the place with Wing Commanders than with these police. They are getting civilian rates of pay. With overtime, these guys are getting one hell of a lot of money.'

I said I had been told that at Farnborough, one police sergeant had been getting £24,000 a year, which was roughly what a Squadron Leader received. One policewoman at Greenham Common, aged twenty, was reported as saying that she liked the work because that year she was on track to making £20,000.

The Air Commodore said he was not surprised to hear it. But an even more important point was what would happen with a heavily civilianized RAF in time of war. There would be a loss of flexibility.

'A serviceman or servicewoman is concerned with serving the Queen, is on duty twenty-four hours a day and does as he or she is told,' said the Air Commodore. 'We have the flexibility to attach or deploy them. We don't have that flexibility with civilians. Even in peacetime, think about being an airman at a heavily civilianized station. You live in a barrack block which is 75 per cent empty and you can't get a game of rugger or cricket. It is a depressing experience.'

In a highly technological service like the present-day RAF was the emphasis on sport and fitness still so necessary? It was not merely a matter of physical fitness, replied the Air Commodore. It was a question that every fighting force had to have esprit de corps: 'You need the feeling that you are playing and fighting with your fellows. That is what may be lost.'

It is not a loss that one would wish to contemplate in time of war, when Support Command would virtually have to stop training, upgrade its communications facilities, transport weapons stocks, recover aircraft from store, send technological manpower to reinforce the front line, bring medical supplies forward, supply men for guarding key posts and provide aircraft for reconnaissance by the police. Esprit de corps would be vital at such a time.

There is another snag to heavy civilianization, which has begun to manifest itself in peacetime. The civilians who are directly employed by the Ministry of Defence on air stations are civil servants. At the time I went to Brampton, the Civil Service was worried because its rates of pay were depressed; a lot of brain power was draining out of the service into the private sector. Brampton had a command research department headed by a civil servant with the equivalent rank of Air Commodore. A civilian member of his staff (with good degree and work experience) had been earning only £12,000 a year. I was not able to interview him at Brampton about his views on this – because he had already left to take a job in private commerce at double the salary, plus a company Volvo.

Even at lower levels it is not always possible to recruit, or retain, civilians with the necessary skills, to work either for a private contractor trying to tender competitively or for the civil service. It may be argued that there are geographical regions in which there is 30 per cent unemployment; but these are possibly not the regions in which the RAF or the private contractor is trying to recruit civilian workers. In an area like Abingdon, where there are plenty of air stations nearby, there are certainly the engineering skills available. But they are already being employed at places like the Cowley motor works, where rates of pay are seductively higher than the Civil Service or the competitive contractor is likely to be able to afford.

These are the sort of riddles the RAF in general and Support Command in particular have to try to solve as they go about their business. That business is intimidatingly vast in the case of Support Command. There are eight flying schools which train 270 aircrew a year as well as twenty-five overseas students and a number of Royal Navy personnel being given elementary flying instruction. The cost of training a fast jet pilot is some £3 million, of which about a third is met by Support Command, the rest being borne by Strike Command which, having the ammunition at its disposal, does most of the advanced combat flying training.

Because the cost of training pilots and navigators is so high, Support Command is trying to find ways of making economies without handing the whole thing over to civilians. It has not been unusual for a station to lose 10 per cent of its pilots at the later stages of training – with the waste of millions of pounds in training time. There are now fervent attempts to identify the failures at the earliest possible moment.

RAF Brampton now has on computer every conceivable statistic about flying training, an area where intelligent changes can save vast sums of money. The statistics are reviewed periodically by the Air Vice Marshal in charge of training and his staff. There is a constant tug of war between traditional assumptions and the need to save money. For twenty-five years, the RAF has used jet aircraft for basic air training, because all the front-line aircraft were jets. But turbo-prop aircraft have developed so much in the past decade that their handling resembles that of jets much more – and represents only a third of the cost of a jet aircraft. So there has been a switch from the Jet Provost to the Tucano aircraft. Some 28 per cent failed on the Jet Provost – an expensive loss to the RAF. But it was discovered that, by giving them an extra sixty-five hours on the comparatively cheap Chipmunk light training aircraft, the failure rate was reduced to 4 per cent.

I asked the Air Vice Marshal in charge of training, who rather resembled an impresario, whether it would not be cheaper still to spend more time on flight simulators on the ground: 'I would like to say use more simulators, and we have increased the use of them. But in the first 250 hours a young man spends in the air, there is little effective substitute for flying time. I would hate to promise that we could knock off a hundred hours on the simulator.'

Why, I asked, did flying cost so much: £1 million to train a transport pilot, £3 million to train a jet fighter pilot (though only £1 million borne by Support Command) and £1.3 million to train a helicopter pilot?

'Because,' said the Air Vice Marshal in charge of training, 'a Tornado pilot will have sixty-five hours on a Chipmunk trainer, ninety hours on a Jet Provost, then another forty-five hours on a Jet Provost if he is doing well, then he will go on to fly the Hawk for sixty-five hours, and *then* he will go to a tactical weapons unit at Strike Command.'

At the time we spoke, the Air Vice Marshal was under remit to reduce the length of time of training by 5 per cent in the current year, and another 5 per cent the following year.

Could this be done without dangerously skimping the essential training? 'The flying done will be exactly the same, but we have looked at ground schools and said, "The man can do such-and-such in two hours instead of three in the ground stages of instruction." We have made the 5 per cent saving this year and we already know how we are going to do it next year. It is easier to monitor the cuts because the stations are on budgets which are the responsibility of the executives on the spot.' Because of ways which had been introduced to spot failures earlier, or to correct failures at lesser expense, some 20 per cent had already been saved on training fighter pilots to squadron standard.

All ranks are encouraged to make suggestions for economies, and are paid for any worthwhile ideas. This is all part of a plan to delegate authority downwards within the Command.

'There is a huge scope for this in the RAF – because there is so much technology there is arguably more scope than there is in the Army,' I was told by the Commander in Chief of Support Command, a spruce fifty-five-year-old Air Marshal who had been knighted after a long career as a pilot. He mentioned a few examples.

At RAF Sealand, near Chester, the biggest aviation centre in the world, military or civilian, many chief technicians had written computer programmes to analyse faults in a particular piece of equipment better than it was being done at the moment.

'He flogs away at his home computer in the evenings and says to his boss, "What about this?" We get scores of these every week. We give them £50, £100, sometimes £1,000. Every week I sign half a dozen certificates authorizing payment. We are trying to make everyone feel they have a part to play.'

At RAF St Athan in Wales, they have a deep maintenance facility for the engines of the Tornado and the Jaguar. Tornado engines, on the front, had an inlet guard of tough alloy, which cost scores of thousands of pounds. They tended to sustain frequent battering and damage; and the makers, Rolls-Royce, had said it was not practicable to repair them. When defective, they had to be replaced. But someone at St Athan had found a way of repairing them which Rolls-Royce agreed was safe.

'This saved us £600,000,' said the Commander in Chief.

Economies on a grander scale can create problems for Support Command, especially for those at the top of the two-thousand-strong community at headquarters. The very *nature* of the Command headquarters, said the Commander in Chief, was an example of slimming down and rationalization to some extent.

Years ago, there were four separate Commands covering the work of the present Support Command. They were Signals Command, Maintenance Command, Technical Training Command and Flying Training Command. They were first merged into two Commands and then into the present one, the group level disappearing and headquarters dealing direct with individual stations.

The Commander in Chief said the Command was unusual in being as large as it was, with 46,000 people; and he thought it was really two Commands co-located rather than one Command: 'The marriage of Signals and Maintenance was natural. The marriage of Technical Training and Flying Training was a natural one. But putting the four together, you finish up with a headquarters which is doing two very different things: there is little read-across from training to the signals side. But really the only difficulty is to me, because I sit on top of two different areas, astride two different disciplines. It makes no difference at all to the way we would operate in war, provided that I and my colleagues can do our job. In peacetime our job is to repair, maintain, supply and train people for their front-line role. In wartime, we would continue to supply and maintain, but to a reduced extent, because deep maintenance over several months would presumably be brought to an end or reduced so we could help to reinforce the front line. If you are going to operate seven days a week, twenty-four hours a day, *and* guard bases and allow for casualties, you would need extra people, and they would be trained through maintenance and training units.'

In time of war, the atmosphere of life at headquarters itself would change greatly (as would life on individual bases, some of which would close down, others of which would switch to war roles). Personnel at Brampton would move into an alternative war headquarters in a secure area to master-mind the sending of Gazelle helicopters to secure runways in Germany, larger aircraft to bolster maritime surveillance and aerial supply and even Chipmunks to provide observation for the military, the police and firemen. With possibly 180,000 Americans coming through to help reinforce the front line, the light aircraft assisting the police are regarded as a necessary precaution against possible subversion and terrorism. The Commander in Chief of Support Command would assume his role of Air Commander Home Defence.

The man at the top of Support Command must plainly be a man of broad experience, with cool nerves in a crisis. I asked the Commander in Chief whether he had a military family background.

The reply was immediate. 'None whatever. My father was in the wine and spirits trade, and we had a family chandling firm. I was called up for National Service in 1950 and stayed in, because they were paying me to fly jet fighters and I thought, "This is something I don't want to give up just yet." I am a fighter pilot who has done six squadron tours, one with strike-attack Canberras and the others on Hunters and fighters. They were anxious to keep people in those days.'

There does seem a tendency to have pilots (whom you would expect to see in *Strike* Command) at or near the top at Support Command, possibly because they are thought to know just what the 'teeth' side of the RAF needs, possibly because it is pilots who are favoured for the top ranks anywhere in the Royal Air Force.

I asked the Air Marshal whether, as a pilot, he had ever imagined himself an Air Marshal or, for that matter, a knight. 'Lord, no! I looked no further than the next tour. There are a few people like Harold Wilson who stand outside No. 10 Downing Street and say, "This is for me!" But I had no thought of anything like that. I was very lucky in that I was very young when I started. I joined when I was eighteen and was on a squadron by the time I was twenty. I was given responsibility early. I was offered a flight command when I was twenty-three and had to be given acting Flight Lieutenant rank in order to take it. Why did I rise? All that I can tell you is that I was idle at school because I could not really see what it was all about. I worked hard in the Air Force not because I was conscious of where it might lead, because I wasn't, but simply because it was so enjoyable and stimulating it was easy to work hard at it.'

Could he have done any other job? He had spent a year as a police cadet, largely with CID, he said. He would not have missed it for anything, because it was the greatest possible education in the facts of life. He didn't come from a frightfully wealthy background, but we all lived sheltered lives in some way or other; and the year in the police had given him an insight into the society of the underprivileged and the criminals which he did not know existed.

He didn't know whether it had benefited him in relation to the RAF; but it had benefited his understanding of what society as a whole was like and what people *could* be like; and he thought that the handling of people was the vital factor for any high officer.

'The only *variable* a commander in the service has got is people,' he said, his enthusiastic ruddy face, young for his age, breaking into an uncomplicated smile. 'The kit is laid down. The pay is laid down. The only thing that makes a commander better than the next bloke is the way he gets the best out of people. A lot of people think the services are frightfully rigid organizations where the Colonel or the Group Captain gives the orders and everyone clicks heels and runs off and does it. Yes, we have got to give instructions and make sure that what has to be done *is* done; but we have to do it in such a way that our people feel they have a role to play and that, if they have an idea or a suggestion, we get to hear about it.'

This idea is no mere whim of a single officer. There are general moves afoot within Support Command to devolve decision-making downwards. One of the most revolutionary originates from the USA: giving an individual Station Commander a set budget, but allowing him to decide *how* he will spend it, and *when* – rather than the traditional system whereby he simply asks for whatever he wants and hopes he gets it.

Budgetization is, on the whole, a new jargon word more unreservedly popular in RAF Support Command than contractorization. About 70 per cent of the 212 Support Command units are on budgets, which gives them an incentive to make economies where required. As the Commander in Chief of the Command put it to me: 'The services never knew what things cost, and, if you have no delegated authority, you have no incentive to do anything about it. The idea is that gradually we will delegate authority downwards. If it works, it will certainly revolutionize things.'

What seems certain is that if anyone in the RAF can benefit from budgetization, it is Support Command, the RAF's housekeeper. It will obviously foster a trend towards promoting officers on the basis of their business skills as well as their military leadership qualities; perhaps *even more* on that basis. If so, it will be a healthier state of affairs on the support end than the 'teeth' end, where fighting qualities need to be the final arbiter of how far a man rises in his profession.

Giving a Station Commander command over his cash as well as his men is also expected to result in having work by civilian contractors done more efficiently and promptly. A Group Captain I spoke to at RAF Brampton said that this

Phantoms at RAF Wattisham are kept for instant use under Quick Reaction Alert procedures. One Phantom of 56 Squadron is seen here outside its hangar.

Germany is in the front line as far as a possible war in Europe is concerned, and RAF Germany has the largest single concentration of RAF personnel and equipment anywhere in the world. The station nearest to East Germany and Warsaw Pact forces is RAF Gütersloh, only a few minutes' flying time from the East–West German border. Here a Chinook and a Puma helicopter, seen with a Scimitar armoured vehicle, are bringing in fresh equipment in the depth of winter.

A Chinook at RAF Gütersloh carries a Rapier firing unit.

A number 3 (F) Squadron Harri[illegible] is even more invisible in its forest hide.

A number IV (A[illegible] Squadron Harri[illegible] GR3 is re-arme[illegible] front of its hardened aircr[illegible] shelter.

The same squadron's Harriers fly ove[illegible] the Möhnesee i[illegible] West Germany.

Laarbruch is further away from the East–
: German border, but there is an air of
ediacy because its Tornados have a
ear capability. The Tornado seen in a sharp
's from 15 Squadron.

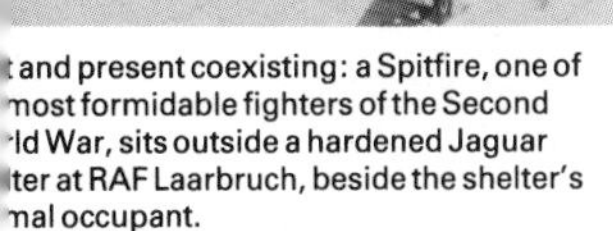

t and present coexisting: a Spitfire, one of
most formidable fighters of the Second
'ld War, sits outside a hardened Jaguar
lter at RAF Laarbruch, beside the shelter's
nal occupant.

nber 14 Squadron Phantoms fly low in
nation over West Germany in the mid-
)s.

A Tornado of 20 Squadron takes off from RAF Laarbruch.

Number 2 (AC) Squadron Jaguar from RAF Laarbruch is seen exercising over Sardinia. The Jaguar is now being replaced with more up-to-date aircraft.

It is easy to think this picture has been turned on its side, but in fact it shows a 92 Squadron Phantom in a vertical climb.

An air defence Tornado launch its missile.

During a station exercise, airmen on stand-by are dressed in equipment offering maximum protection, though often hot and uncomfortable to wear.

Helicopters are invaluable for delivering supplies. Here an 18 Squadron Chinook (*above*) is seen with an underslung load above the German countryside.

An emergency evacuation of a squadron headquarters at RAF Brüggen, We Germany, in January 1987, following an 'air raid' in a station exercise.

The great distance between Britain and the Falkland Islands – some eight thousand miles – has been conquered with the aid of refuelling in the air. Here a fully armed Phantom from the Falklands is being refuelled from a Hercules tanker (*left*).

At Mount Kent in the Falklands domes cover vital radar equipment (*below*).

Life for the RAF in Belize in Central America is tropical but taxing, especially when practising 'dips and dunks' on one of the keys (sandy islands) around the small country (*left*).

ere are some advantages to the tropical life of at was once part of the British Empire, as this ing member of the RAF Regiment (*above*) uld agree as he relaxes – though with his rifle at side – by the river near the road between itary camp and Belize City.

A wing is changed (*above*), with help from the British Army, in Belize.

A man of 1 Squadron the RAF Regiment takes part in an exercise at RAF Gütersloh (*left*), with the heavy armour of the British Army in the background.

RAF Regiment personnel (*above left*) prepare a Rapier fire unit, which sends a Rapier (*above right*) off towards its mark.

Training in the RAF Regiment takes place at the depot, RAF Catterick, in an arduous way no less demanding than that for the British Army, whether it is climbing over forbidding walls (*left*) or worming one's way under barbed wire (*above*).

RAF Brize Norton runs the 'world bus service' for all three armed services and also provides tankers which refuel aircraft in the air (*above*), greatly increasing the flexibility of the RAF's presence anywhere in the world. (*Left*) A VC10 transport aircraft takes off from Ascension Island on its return from the Falkland Islands to Brize Norton.

A VC10 tanker undergoes a pre-flight check at RAF Brize Norton.

A VC10 of 101 Squadron Brize Norton – a tanker supplying two RAF Jaguars over the North Sea.

The publicity value of the spectacular Hawks of the famous Red Arrows is considerable. They are shown in formation (*above left*) and manoeuvring close to one another in a procedure called the Twizzle (*above right*).

was already happening there. Under the old system, when a job had to be done to married quarters, the civilian contractor knew that the Station Commander had no authority to pay out the money, and would tend to arrive a week or a fortnight later – or when he felt like it. Under the new system, when the Station Commander or his deputies would be signing the cheque, they turned up to get the job done the very next day. And, because they knew the Station Commander could query the bill, they would be less inclined to try to pad it, which would save the RAF money.

One officer at Brampton told me that the fact the pilots left the service early cost the RAF a fortune. 'If you could keep a pilot to the age of fifty-five, you could give each pilot a million pounds gratuity and still save the defence budget £100 million a year. Every chap who leaves before full retirement date costs us money.'

The RAF puts into squadron service 165 pilots a year, but the average time they stay in the RAF is sixteen years. It is obvious that, with the £3 million cost of training a fast-jet pilot, if a pilot stayed for double that time, it would effectively halve the cost of training. Some officers are incredulous that the RAF is prepared to pay a pilot to leave. At thirty-eight, he can decide to leave with a gratuity and a pension. Since, by this time, most pilots are conscious of fast-rising house prices and frightened they will never buy a house of their own if they do not do it then, many decide to take advantage of the profitable exit. There is a strong feeling in the housekeeping command of the RAF that there would be some sense in reversing the procedure, so that it paid a pilot to stay in rather than go out.

But it is also thought that there could be economies at less exalted levels. Seven hundred airmen enter the RAF every year under the government's Youth Training Scheme, and three-quarters of them stay. This is regarded as a great economic success both for the government and the RAF, which takes 10 per cent of all youths who enter the scheme – unlike the British Army, which has found only a small number of recruits suitable for its own purposes.

The use of women wherever possible leaves men free for the 'teeth' roles. More and more roles in the RAF can be filled by either a man or a woman. 'You will never know until the telephone is answered whether the person you are calling is a man or a woman,' one officer told me.

'And sometimes not even then,' put in a junior officer.

There are women cadet pilots in the University Air Squadrons, which provide an increasing number of candidates of both sexes to the RAF. But there are economic as well as psychological reasons advanced for why it would not be expedient to have women trained to be pilots in the RAF. Though the average male pilot stays fifteen years, women in the RAF stay for an average of only about six years. If they were trained as pilots, the cost would effectively be double that for men.

I asked the Commander in Chief his personal view on introducing female pilots. 'I happen to think that a proportion of female pilots would be good thing. Where you have women in a group, general standards rise. Men can be stupid and macho, but if there is a woman instructor you would try that much harder to do the role better than she does. But there are snags. The Americans, who have tried it, are now saying it doesn't always work very well. Once the women had proved they *could* do as well as men, it became, "Do I *have* to get into that smelly flying suit?" I think we would have to run out of men to make it economically sound. I can't see it coming for ten years.'

It could take even longer than that if the economy continues to be the dominant refrain in Support Command. In a way this refrain has been invigorating, causing all ranks to use their brains to think up ways of rationalizing their jobs. 'It is wrong to think we are a totally technologically based service,' said the Commander in Chief. 'This is as much a *people* organization as the Army. The difference is that we fight with kit, not with soldiers. But the more we look after people, the more they have a way of telling us what they feel and think, the better. If you go around the maintenance side, and you look at the enormous improvements and savings that have been made, you will be told that very many of them have been made on the initiative of the shop-floor. One man saved £20,000 a year on cutting grass – there is a lot of grass to be cut on most stations.'

No doubt the cutting of grass is important. No doubt saving money is important. No doubt there are areas where making the Royal Air Force more like the RAF Plc through the use of civilian contractors can save money. But it would be odd if the one determined British government that would not have simply intoned fine words over the Falklands and then run away goes into history, also, as the government that compromised our future fighting ability in an internationally crucial war by confusing shopkeeping economics with the imperatives of success in war. It would be even more odd if the position had to be corrected by political opponents who are often thought of as no friends of the military.

But at least one Royal Air Force activity is *highly* unlikely to be tackled by the most intrepid civilian contractor.

9

THREE-DIMENSIONAL CHESS

'The Captain of an air-to-air refuelling squadron, which was refuelling Vulcan bombers on their way to make a Falklands raid, told his crew, "If we don't give the Vulcans more of our fuel, they won't be able to make the raid. But if we *do*, we won't be able to get back home." And, to a man, all the men on that squadron shouted rousingly, "Don't give them the bloody fuel!" . . . That's a *joke*.'

– Pilot of 101 Squadron (a Wing Commander), on refuelling duties at RAF Brize Norton.

'You can't have a wizard-prang mentality if you are air-transporting passengers. You have to ask yourself, "Would I want my wife and children in this plane the way I am flying it?" You can't have flashiness in this game, it consumes a lot of energy. What you *must* have is infinite flexibility and a can-do attitude.'

– Pilot of 10 Squadron on air transport duties at Brize Norton (a Flight Lieutenant).

Couldn't running the air transport service of the RAF, sometimes colloquially known as 'the world bus service' seem a rather humdrum job for a serious fighting man? The veteran Flight Lieutenant laid a proselytizing forefinger firmly against my chest.

'I will tell you a tale about how "humdrum" air transport can be,' he said. And did.

It was in the 1960s, and he was on his first assignment

in air transport, as a navigator. He breezed into the Operations Room, and asked the pilot where they were going.

'I can't tell you,' replied the pilot.

'In that case I'm going home! What do you mean, you can't tell me?'

The only answer to this explosion of hurt pride was a stiff aviator's upper lip and perfect silence. Only when they were walking towards the old Beverley aircraft did the pilot relent to the extent of saying he had sealed secret orders he could not open until they were airborne.

The orders opened, they turned north from their initial route and flew over a Middle Eastern desert, picking out a certain formation of rocks, sighting a certain single tree and then landing between two sand dunes on a strip of characterless sand which, they were assured, was firm enough to land a Beverley on. They taxied to a halt and waited for over an hour in the middle of nowhere as the aircraft got hotter and hotter.

Then what appeared to be an Arab camel train appeared over the horizon, attended by bearers of swords and rifles. Neither pilot nor navigator knew what to expect. They had one gun in the aircraft. It was possessed by the co-pilot so that he could shoot all the crew rather than allow them to be taken prisoner and mutilated: a possibility only slightly reduced by the habit, at that time, of carrying a card in the native language saying that a certain sum of money would be paid to anyone returning the airmen *un*-mutilated.

The co-pilot held his fire. The camel train, ferocious-looking but peaceable, loaded up some sacks on to the Beverley, the contents of which were not described. The young navigator did not breathe freely until they had taken off again and completed the return journey.

'So don't talk to me about the transport service being humdrum,' said the Flight Lieutenant, now in a training job at RAF Brize Norton near Oxford, the station which handles much of the RAF's transport for personnel, as well as much of the RAF's air-to-air refuelling activities – both vital links in the chain of the worldwide RAF presence protecting British interests.

Of course the air-to-air refuelling people at Brize Norton agree that transport isn't *humdrum*, but they then add that, of course it *is* more 'structured' (a less offensive word) than their *own* lives, which really *are* lived at the sharp end. I had not been at Brize Norton long before I saw what they meant. A number of VC10s – those delightful British-built aircraft so familiar to users of British Airways – landed well in time for tea. They had just spent the day refuelling in the air 617 Squadron (the Dambusters) on its way to some training exercises at Goose Bay in Labrador, quite near the Arctic Circle of Canada. It entailed a journey of some five thousand miles.

The Squadron Leader of 101 Squadron, the Brize Norton-based unit, still wearing his flying overalls, was well into some paperwork when I arrived at his office. A tall, red-faced Northern Irishman who had entered the RAF via Queen's University, Belfast, he told me his day had started at seven in the morning when he, and the crew of five of his squadron of nine VC10s, had had a meteorological briefing for the Atlantic before escorting nine Tornado fighters. It had progressed with the Tornados being refuelled in three separate waves, while the Squadron Leader himself did a 'whirler'.

I looked puzzled. This, said the Squadron Leader, was when you refuelled in the air the first wave of the aircraft you were fuelling, broke off, got *yourself* refuelled in the air and then turned to refuel the last wave.

The truth is that it takes even an interested civilian some time to understand the three-dimensional chess of the strange craft of air-to-air refuelling – a craft that could be vital in war, since it is the means of getting aircraft to otherwise too-distant parts of the world, and of keeping in the air enough aircraft to protect the United Kingdom from any attacker, even if their own home airfields have been wrecked.

It is a spectacular craft practised with a strong nerve, a

cool head and a two-and-a-half-inch-diameter fuel pipe capable of pumping one tonne of fuel per minute from the trailing pipe of the feeder plane into a nozzle facing forwards on the receiving aircraft.

'I am now forty-three, and I have been in the business of tankers since 1969, first flying Victors and then VC10s,' said the large Squadron Leader after a handshake which almost qualified me for the sick bay. 'I have never wanted to do anything else. I enjoy the flight refuelling business immensely. I am a great believer in its usefulness. It is a real job, not *practising* for something else. Fighters are going to continue to fly, or crash, according to what you do. In a lean RAF – and it is getting leaner and leaner, with us being asked to do the same job with fewer people – the one key issue in air power is flexibility. You must get the very most out of the machine.'

Just so. An air raid that would have required four or five squadrons of the old Second World War Lancaster bombers to provide an acceptable percentage chance of success can now be accomplished with four Tornados. But the Tornados could not reach every target and get back home without being refuelled; and, in war conditions, this means being refuelled in the air.

The process can have mechanical difficulties. On one occasion the Squadron Leader, while still flying Victors, encountered one of the most serious: seeing his receiving aircraft getting a serious malfunction in the middle of the Atlantic, right at the point of no return, when there is just as far to go as to come back. It was a Jaguar fighter, one of whose engines had cut out when it was flying over cloud directly underneath. It was shortly to be refuelled. Instead, it disappeared from the sight of the Victor as it plunged, underpowered, through the clouds.

The Squadron Leader was supposed to be on his own, refuelling this and another Jaguar. It was Friday afternoon, and he was due to be best man at a wedding on the Saturday morning. The wedding might have been delayed if it had not happened that another VC10 pilot had been due to fly the Atlantic and decided to keep the other company. The Squadron Leader asked this VC10 to dive through the cloud. When the Jaguar was at last found, the other pilot escorted him to Iceland where he made an emergency landing for repairs and refuelling.

The Squadron Leader got to the wedding on time.

'If I had been on my own, I would have had to turn the whole formation to Iceland in order to keep with him,' said the Squadron Leader. 'And to think I was a Whitehall warrior, as it happened, during the Falklands war!'

Many problems faced by the RAF's air-to-air tankers are not merely mechanical; they are a subtle blend of the mechanical, the mathematical and the moral. This combination applied especially in the Falklands fighting.

A Squadron Leader navigator at Brize Norton, who was in the Falklands campaign – based on the staging post at Ascension Island, halfway to the Falklands – remembered the first of the Vulcan raids on Argentinian targets. At this time, the amount of fuel used by the Vulcan bomber was based on peacetime calculations, which were concerned with the most economical cruising speed, and suchlike matters.

The navigator remembered the unfortunate consequences. 'Essentially we had forgotten how much fuel a Vulcan used. Flying a Victor tanker, which was doing the air-to-air refuelling, the pilot had to make the appalling decision whether to give the Vulcan a chance to do the raid, by giving away more fuel than he could afford (*all* the fuel on a tanker can be used by the tanker itself or given away), which would leave himself without enough fuel to get back to Ascension Island. The pilot knew, when he turned and came back to the Vulcan by himself, that he had insufficient fuel, by about one hour, to get back to Ascension. They maintained radio silence until they were about two hours out from Ascension Island, when they heard the code word from the Vulcan that its bombs had hit – if they had said anything *before* then on the radio it might, by implication, have told the Argentinians what was going on, and spoiled the raid. After they heard the code word, they opened up the radio and called for someone else to meet them with more fuel. The pilot was given the Air Force Cross.'

Those on the ground at Ascension on that occasion also had their moral problems. Partly because the Vulcans had been using more fuel than expected, and therefore some had had to be unexpectedly refuelled, there were half a dozen aircraft all wanting priority landing clearance on Ascension at the same time. Should the controllers on the ground allow one aircraft down before its predecessor had taxied clear of the runway, thus risking a pile-up between two aircraft? Or should they deny clearance until the safety rules had been met, thus risking seeing the second aircraft crash with an empty fuel tank? Some of the aircraft hadn't sufficient fuel to make even one more circuit of the airfield. Ground control decided to get all the planes down as quickly as possible, and managed it without collisions.

The received wisdom on this point is that it is better for an aircraft to be shot down in combat – when it has some chance of doing damage to the enemy – than tamely to crash because of lack of fuel. There is a strong pressure, therefore, on tanker crews to give away every kilogram of fuel they can spare from their own tanks – larger in the case of the newer TriStar tankers which, at the time of my visit, were gradually increasing in numbers at Brize Norton.

The pressure is sometimes misunderstood or resented by younger officers. One officer at Brize Norton remembered when he was a young co-pilot of a Victor tanker. They took off from Gan, in the Indian Ocean, to escort some fighters to Singapore. The Victor had to be launched in the early morning, before the sun became really hot, because a hot atmosphere meant more fuel was needed for the same forward thrust. They carried just sufficient fuel for a safe take-off and the provision of enough fuel to the fighters to get them safely to Singapore. The pilot used all the runway for the take-off in order to save the fuel that would have been necessary for a quicker, steeper climb.

When the refuelling was under way, some bad weather

started rolling in from Singapore. The experienced pilot knew at once that the fighters would need more fuel than had been planned to get them to Singapore. The young co-pilot worked on his charts and told the pilot, 'We *can't* give them any more fuel. We will run out ourselves on the way back.' To which the veteran pilot replied simply, 'Give them some more.'

The pilot made a gliding approach back to Gan, closing the throttle and letting the aircraft descend gradually. He applied power only when the aircraft was six hundred feet up and he landed with what the now veteran co-pilot described as an 'indecently small' amount of fuel left in their tanks – much less than the technical minimum of eight thousand pounds.

The rather shaken young co-pilot got the point when news came through of what had happened to the fighters trying to get into Singapore. Because of the bad weather, one of the fighter pilots missed the airfield at his first approach, had to abort a second time and then had to divert to another airfield. Only the fact that he had been given the extra fuel enabled him to do this. It had perhaps saved his life; certainly it had saved the life of his aircraft.

In the Falklands campaign, tankers landed without anything like the wartime minimum (four thousand pounds) in their tanks. One had no fuel registering on its gauges at all. In peacetime, the hazard lies more often in the fact that fighters watching Russian aircraft to the north of the British Isles have to stay up for longer than they bargained for, and must be refuelled in the air unless they are to lose sight of the Russians, who will possibly be using the so-called Bear, a large propeller-driven bomber-cum-spyplane which is capable of a surprising turn of speed and manoeuvrability.

In such cases, a VC10 from Brize Norton may have to refuel, land, take aboard more fuel and head northwards again.

Expert ground crew say that the average time for refuelling a VC10 is one tonne a minute, plus twenty minutes –

and point out that nothing else (such as mechanical check-ups) can be done to the aircraft while it is being refuelled. Since the aircraft will already have its safety minimum of fuel aboard, it is expected that VC10s can be turned round very quickly.

If a tanker goes to refuel a fighter and finds that the fighter is facing more than one Russian aircraft, it will stay in the area to watch the second Russian aircraft.

But the main problem of the tanker crews' life is not seeing Russian aircraft; it is *failing* to see the aircraft they have been sent up to refuel. It is comparatively easy to *fail* to see when, in mid-Atlantic, without the aid of radar and in thick cloud, the tanker pilot tries to find the aircraft he is supposed to refuel. 'To see the aircraft at all,' said one 101 Squadron pilot, 'you may well have to climb well above the aircraft and the cloud. This may mean that it is impossible to refuel. A Lightning will refuel at up to forty thousand feet quite happily, but a Buccaneer or a Tornado needs a lower level. You may have to make contact with the aircraft by both of you getting up above the clouds, and then you have to take him down through the clouds to a level where you can refuel him, keeping contact all the time so you don't lose him.'

There is what is called a Standing Operating Procedure. Under these rules, once visual contact is established, the pilot of the tanker becomes formation leader; the recipient aircraft is under his command. In wartime, the procedure provides for the recipient to come astern and connect his nozzle to the conical 'basket' containing the female junction, the whole being trailed from the tanker in its own slipstream. In peacetime, the pilots communicate by voice via radio. 'Clear to join on the port,' a tanker pilot will tell one recipient and perhaps a few minutes later, 'Clear to join on starboard' to another recipient: VC10s can refuel two aircraft at the same time if necessary.

Though an aircraft may have positioned itself correctly behind a tanker, the refuelling may not start at once. It is up to the tanker pilot to choose the best moment, taking into account the weather, the position of other aircraft (perhaps Eastern bloc) and the urgency of the need for fuel. It is not uncommon for aircraft to fly within a few feet of one another for a quarter of an hour, a procedure requiring extreme steadiness of eye and nerve.

I asked how much of the procedure was laid down, how much had to depend on the intuition of the moment. There was marked disagreement in the answers. One pilot said that 98 per cent was laid down in the Standing Operating Procedure, and that it was only when things went awry that you had to use intuition. But a navigator said that, in his view, it was 98 per cent intuition, in that intuition was required to know how to apply the standard procedure to the conditions actually obtaining at any particular moment.

At the time I visited Brize Norton, the longest trek was out to the Falklands, which required the whole squadron of nine VC10s to go via Ascension – though, with the gradual introduction of the TriStars, a direct journey looked possible. In peacetime, aircraft are never put into a position where they cannot divert to a safe airfield if they run out of fuel, though in war they would find the nearest ship and then eject from the aircraft somewhere near it.

Making sure that, in peacetime, they never get to the point of no return is achieved by using what is called the Bracket System. The map is divided up so that sections which are said to be critical are clearly shown. An aircraft must have sufficient fuel aboard before it can enter a critical area; if he hasn't, the pilot must divert to an indicated airfield. What effect the arrival of the TriStars of 216 Squadron at Brize Norton will have on this Bracket System only time will tell. But the fact that these aircraft are so much larger (with room, in the passenger mode, for 250 passengers instead of only 130 or so) and have 1980s technology rather than the technology of the 1960s, makes for optimism.

Three TriStars were bought from British Airways and two from the American Pan-Am company, in the first instance; some were already in operation during my visit. One of the pilots, formerly a test pilot for the Boscombe Down Aircraft and Armament Experimental Establishment, said the TriStars were beautifully quiet and, for that reason, were welcomed at airfields around the world, where the noise of the VC10 would be unpopular. In the passenger role, they were popular because they had in-flight entertainments and the pleasant environment given by the wide body. The American aircraft had not been so well looked after aesthetically as the British Airways ones – the seats were scuffed and the windows were dirty – but they flew well, I was told.

I asked this experienced pilot whether the TriStars had any disadvantages. They might, he said, have difficulty, if fully loaded, in taking off from airfields with only a 6,000-foot runway – but most airfields had at least 7,500, and an airport like Heathrow some 14,000 feet. Also, for technical reasons concerned with the length of the fuselage, the speed of decisions that had to be made on take-off was greater than in the case of the VC10s.

The pilot thought that, overall, it must make sense to go for the bigger of the two aircraft, especially the multi-role tanker-cargo-passenger TriStar being introduced later, which would have huge front doors, through which Land Rovers could be driven.

But when I went to Brize Norton, the tanker and passenger roles were still distinct. The passenger side was being handled by 10 Squadron, in whose Squadron Operations Room I started watching the preparation of a typical flight (if there *is* such a thing) for the passengers. This was Flight 2589, which was destined (though its purpose could be changed at a minute's notice) to take ground support personnel to Goose Bay, Labrador, to join personnel sent there the previous day for an exercise.

The Squadron Leader who was in charge of the squadron operations that day, a compact, sharp-featured and alert-looking man, told me that they had been dealing with Flight 2589, in some shape or form, for the past two months.

Some six weeks before, the squadron would have received what was called a Monthly Tasking document, setting out all the things they would be expected to do: a mixture settled at meetings of Whitehall warriors long *before* that, but roughly consisting of 60 per cent British Army passengers, 15 per cent Royal Navy and 25 per cent RAF (for the RAF is world cab service to all three services).

After the Monthly Tasking document, Squadron Operations Room would receive an operations order called a Transop which would give all the necessary details – too legion to mention here – about Flight 2589, from the time its crew were to be called at their homes to the time the aircraft was to touch down again at Brize Norton.

But planning flights for the forces is not so straightforward as planning them for holiday-making civilians at Heathrow, Gatwick or Luton. Just like the sharper end of air-to-air refuelling, this job has its own elements of three-dimensional chess. It is made necessary by the fact that, though speed is vital and the distances to be covered are long, the working day of aircrew is limited to sixteen hours, including the two hours before the flight, which are reckoned to be flight preparation time.

The implications of this are not immediately obvious to a civilian. They became more and more obvious as I watched the progress of Flight 2589. The men to be taken over to Goose Bay were actually based at RAF Marham, in North Norfolk, twenty minutes' flying time from Brize Norton. They would have to be picked up at Marham and then flown to Goose Bay, some five hours' flying time.

So one crew could simply take the aircraft over to Goose Bay, right? *Wrong.* If they did, they could not fly again until they had enjoyed their proper rest period – and the top brass required the aircraft to be back at Brize Norton as soon as possible. So a slip crew had to be introduced – a crew put aboard an aircraft not necessarily to fly it but to take it (or another aircraft) over at a later stage. The idea was that the slip crew would fly the aircraft to Marham and stay there until the aircraft came back to Marham on its return flight some twenty-four hours later. The aircraft itself would remain overnight at Marham, and then the other crew would take it to Goose Bay *and* back in the same working day, which was just possible within the sixteen-hour duty limit.

They say that infinite flexibility is the magic formula for transport men, and I began to see why: one set of individuals had to fly a total of forty minutes out of some thirty-six hours, spending the rest of the time at Marham, away from home and fiddling their thumbs, while another set of individuals had to fly the Atlantic and back in the same day, arriving home exhausted.

Sometimes more than one slip crew is necessary. For an exercise in Nairobi, slip crews were flown into Cyprus and Bahrain. Officers at Brize Norton remember this expedient extremely well because, just before the flight was due to take off for Nairobi with British Army passengers, a top-ranking officer came into Squadron Operations Room and said, 'Gentlemen, the exercise was cancelled a month ago by the Army.'

Dealing with the unexpected is the very essence of the job. One officer I spoke to, a pilot, said he had been due to fly out to Calgary that day, but the flight had suddenly been called off: 'I have stopped asking myself *why* I'm not flying to Calgary. You're too busy to bother. Ours not to reason why.'

Another officer pointed out that flights often had to be cancelled five minutes before they were due to leave, for a number of possible reasons. Perhaps a political problem had sprung up in the country you were going to or over-flying. Perhaps there was a civilian air traffic control strike, so that the RAF could no longer, as usual, depend on the civilian air control structure.

The Squadron Leader in charge of the Squadron Operations Room pointed out: 'When we, for instance, go out to Cyprus and back in one day, it is close to our maximum working day. An air traffic strike tends to cause lengthy delays as opposed to cancellations; but a lengthy delay means you have gone outside the crew duty day, and the aircraft would have to stop somewhere.' His colleague, a navigator, added: 'We have a particular problem. Military aircraft aren't allowed to over-fly Swiss air space. We have virtually got to fly over France if we are going to Africa – so we are particularly vulnerable to French air traffic strikes.'

The Squadron Leader in charge told me that the Operations staff had made out the crew list and the route the aircraft was going to fly; but the Captain, arriving two hours before the flight was due to take off, would make out his own maps and charts, and specify how much fuel he needed for the journey.

The constitution of the crew would have been vetted in advance, however, by the Squadron Operations Room. Inexperienced co-pilots would not be flown with people only just promoted to Captain. If a man had just been converted from co-pilot to Captain, Ops would make sure he had an experienced crew with him. Ops would ask questions about the airfield that was to be flown into, and then the Captain would sign the authorization sheet that signified he was happy with the preparations.

Flight 2589 was due to take off at 2 p.m. local time. Members of the crew who lived locally in quarters had been picked up two or three hours before this. The ones who had bought their own homes had to make their own way in on time, by means of alarm clock and car or motor cycle. The worst thing that could happen, said the Ops staff, was when a crew member failed to turn up, or when an aircraft went unserviceable on an airfield on the other side of the world.

'Especially,' put in the Squadron Leader in charge, 'when it happens at two o'clock on a Sunday afternoon. You have got to get another aircraft out there, and it throws the whole job pattern into disarray. To Calgary, for instance, one aircraft involves three crews. You put slip crews into Germany and Calgary. You have an aircraft going between places; one crew on the ground at Gütersloh, one at Calgary and the other flying in between. There are usually only two aircraft for this particular lift, and that means six crews. If anything breaks, the repercussions are rather severe. You need to find *another* crew to send there. If it happens at two o'clock on a Sunday afternoon, it means the top brass has to come down here to pick the bones of the problem. But one thing we can say: we have been running Victor tankers from here since 1966, and have never had an accident.'

At the *Station* Operations Room, in a separate new brick-built block at Brize Norton, they have an overview of all flights going in and out of Brize Norton: tankers and taxis. The Squadron Leader in charge of this domain, a gentle-faced man with strong views, told me drily that *passengers* were the bane of his life.

'They never turn up on time,' he lamented. 'They are usually senior officers who think there are good reasons why they shouldn't be there on time, and why they should be allowed to walk on to the aircraft first. The trouble is that people, spoiled by the civil airlines, are used to flying silly. They turn up two minutes before the doors close. Whereas here, we check in one and a half hours before take-off for the VC10s and two hours for the TriStars.'

Told they couldn't fly when turning up only a few minutes before take-off, some servicemen demanded angrily, 'What the hell is the RAF for, if they can't fly us?' But the Squadron Leader in charge of the Station Operations Room pointed out that Brize Norton did not have the sort of computerized facilities for passenger handling which were used at Heathrow, and were not likely to get it. 'We are still in the steam age. We haven't got the money to bring it up to the twentieth century; and our security is slower.'

Inside the Station Operations Room were a number of long desks, facing a giant blackboard, on which the individual flight information was shown on vertical panels. The Co-ordinator (the Squadron Leader's deputy) sat in the middle of the middle bench, with a row of movements experts on his left, and a group of engineer experts to his right. The blackboard they were watching was entirely manually controlled. Flight information, in words and figures, was put, character by character, on to the pegboard panels, by hand. Cut-out red arrows on cardboard were pegged and re-pegged to the board – with aircraft rivets or matchsticks to show at what stage the aircraft now was (whether it was still at Brize Norton, whether it was on the airstrip or at one of its perhaps several destinations).

I questioned the use of the matchsticks. The Squadron Leader was unapologetic. 'Primitive, yes, but it works – and would go on working in war. We have thought about computers. But if you go to war, someone is going to muck up your power supply. Sometimes very sophisticated things pack up. We have got to stay mechanical, *not* computer.'

Talking about the details of another flight in the same Ops room was its pilot. His dilemma showed clearly enough the constraints the passenger service is faced with in peacetime. The pilot had been due to fly to Dulles international airport, Washington, and then on to Grand Cayman in the Cayman Islands. He had arrived for duty at six that morning, which meant that his working day would run out at ten that evening. And there was a mechanical fault on his plane, which held up his departure. If there had been no subsequent need for the aircraft, it might have been possible for the pilot and aircraft to spend overnight in Washington, and then go on to Grand Cayman. As there *was* a pressure, a slip crew had to be drummed up at short notice and put aboard the aircraft, so they could take over at the end of the original pilot's working day.

Just as the tanker pilots have a way of overriding the safety device which prevents them giving away too much of their own fuel, so the pilot of a passenger plane can elect to go over his shift time. But in the case of neither the tanker pilot nor the passenger flight pilot are the respective self-sacrifices encouraged. In the case of the passenger pilot, the safety factor is paramount: a tired pilot could lead to everyone on board losing their lives.

Once in the air, the average pilot has to face choppings and changings that demand his fullest and brightest attention. Just before my visit to Brize Norton, a VC10 had been detailed to fly across Burma to reach its destination in India.

The Burmese directed the pilot to change course, because its original route would have taken the aircraft too near a VIP about whom they were jittery. This meant that the pilot had to try to fly into India on a different route. This was too much for Indian bureaucracy, which insisted on the status quo. The pilot, doing five hundred miles an hour, was due to reach the Indian border in one hour's time. He couldn't go back because of the VIP; no aircraft can stop dead in the air, except the Harrier; and to enter Indian air space at an unauthorized point could provoke an international incident.

Station Operations were up to the situation. They raided their contacts books for the necessary telephone numbers, rang the British Embassy in Delhi and got them to bring diplomatic pressure in favour of authorizing a fresh entry point to Indian air space. 'All we had to say was, "Get moving!"' said the Squadron Leader in charge of Station Operations.

The man responsible for the passengers from the time they set foot in Brize Norton is the Movements Officer, working chiefly between the booking hall and the tarmac. As people checked in for Flight 2589, he, a Flight Lieutenant, told me that the system was only partly computer-backed, although it was being upgraded; that he could cope with passengers after the proper check-in times (at the end of the runway if the need was urgent enough); and that the pilot's word was law when it came to the security of his aircraft.

Quite recently there had been a case, in Belfast, of the aircrew's worst worry: the head-count of people on the aircraft just before take-off being one short of the listed number. The obvious possibility, in such a case, is that the missing person has left a bomb on the plane. At Belfast, it was rapidly established that the missing person was a girl, with a baby, who was still standing on the tarmac. She was nervous of flying and angry that her husband had tried to make her fly. As she reluctantly reached the bottom of the entry steps, she handed her baby to the coaxing Movements Officer, so that she could hit her husband. There was a general fracas. The pilot got out of his seat and told the Movements Officer to get everyone off the plane and not bring them back until the aircraft had been thoroughly searched for bombs. Some passengers thought his attitude was ludicrous, and told him so.

'Better ludicrous than dead,' said the unmoved pilot. 'Most people get the point that it is better to be delayed than blown up.'

Before the passengers got aboard Flight 2589, the provisions had to be there. Providing food for the RAF's passengers is a gigantic operation, which draws on the skill of the RAF's own cooks, and on the private sector. It is the responsibility of the Catering Officer in charge of Gateway House, the large grey brick-built block several hundred yards from the main airport buildings, which accommodates personnel and families when they are to take an early morning flight or when they return too late at night to go on to their destinations.

The Flight Lieutenant who filled this role was thirty-eight and belonged to a family renowned in the catering field. He had previously been in the family business, but said he preferred working for the RAF. He was checking on the loading-up of two large metal boxes, each containing eighty meals. These were to be taken to Marham, put into cold storage there and put back on the aircraft the following morning, to be taken to Goose Bay.

His dual role as hotel manager *and* provider of in-flight meals was obviously no job for a lazy man. Gateway House is the only custom-built transit centre the RAF has. It can house 350 people in 170 twin rooms. On the occasions I have stayed there I have found it as efficient and comfortable as many an hotel, but the Flight Lieutenant said I really mustn't speak of it as an *hotel*.

'I don't like using the word "hotel", because we are not established, either in terms of staff or equipment, to provide what people would understand by "hotel" service,' he told me. 'We *look* like an hotel, but we are an air transit centre – we are not here to provide room service.'

The diligent caterer, who had been fourteen years in civilian catering before getting out of it with some relief, examined his remark and decided he didn't entirely like it. He added quickly: 'But what we are currently doing is to increase the number of leisure facilities, to make it more comfortable for passengers. We put in a multi-gym, a children's playroom, an outside adventure play fort. We are putting in a badminton court. We are trying to update the flight information boards, and the eight VIP suites have been completely redecorated.'

The non-hotel is free for servicemen on duty, but so-called 'indulgees' (either servicemen off duty or their families) pay – or paid at the time I was there – only £6 to £18 for a night's accommodation, plus 97p for a main meal and 62p for breakfast. It is a bargain, except perhaps for the staff, who have to provision 3,500 flights a year as well as run Gateway House. This means that some 200,000 passenger meals and about 35,000 meals for crews are provided in the course of a year.

Some of the food being taken aboard Flight 2589 had been bought from commercial contractors. This was the food destined for the passengers. The food for the *crew* was prepared entirely by the RAF's own cooks. This routine appears to be based on the theory that, if anyone outside tampers with the food in any way, it will do less harm if the passengers are poisoned than if the pilot is.

In the kitchens, I was able to see examples of both the bought-in food and the RAF's own. I asked the Catering Officer which food was best. I did not have to wait long for the answer. 'The RAF's, without question.'

Which, in a sense, was *why* the Catering Officer was no longer in civilian catering: 'The object of this exercise is to provide food for people. The object of civilian catering is to make money – providing the food is the means to that end. In private catering, you have to put up with poor staff and the rat race – cheapen this and cheapen that! There is no rat race here – other than straightforwardly for promotion, I suppose, and that certainly can't be achieved by stepping on someone else. Here I like the life-style and the work. Put these together, and you are entirely happy at the beginning and the end of the working day. I was pleasantly surprised, after fourteen years in civilian catering, when I realized I was still not too old to join the RAF, which had been a previous ambition before I was pressed to go into the family business.'

Of course, in private catering he did not have one constraint: having to produce three meals a day for £1 17p.

When the Catering Officer told me he was in the course of conducting a survey on what aircrew thought of the RAF food, I thought (with the £1 17p for three meals at the forefront of my mind) that this might constitute masochism.

But no, it appeared the early responses to the survey had been analytical rather than critical. The Catering Officer had asked for comments on all the twenty-two main courses which are prepared during any one day, and on the ancillary items. Despite the trend towards healthier eating, the rainbow trout and salad were not popular with aircrew: they found bony fish difficult to eat when in an aircraft, especially when at the controls. For the same reason, steaks were unpopular – but here the Catering Officer thought tenderizing might be the answer. Curries were popular, again probably for the same basic reason. Most people left their cracker biscuits and puddings, and did not like the quality of the bread rolls. The Catering Officer said he was now writing a paper recommending what he thought were necessary changes.

Apparently the passengers eating contract food would have to make do without a survey. But then they do not have to eat aloft *regularly*, as crew members do.

After Flight 2589, freshly provisioned, had taken off five minutes *early* (RAF transport flights do not wait till the appointed time if there is no reason to do so) one or two pilots told me they were, in a way, sorry that I had seen such a trouble-free operation, rather than witnessing the things that can sometimes happen in the strictly *non*-humdrum RAF transport world.

There was the Brize Norton pilot who had been bussed to RAF Coltishall in Norfolk to stay overnight, in order to meet and fly back to Brize Norton a flight from the USA. The aircraft developed a fault in Los Angeles, and was not even seen by the pilot, who was bussed back to Brize Norton the next day, feeling like the Grand Old Duke of York who had ten thousand men, and marched them up to the top of the hill, and marched them down again.

There was the pilot who remembered how, as a young co-pilot expected to handle the financial affairs of the travelling crew, he had been grounded in Washington for a week because the aircraft had to be repaired, had run out of cash and had had to go to the British Embassy to prise some more money out of them. 'And diplomats can be idle buggers, who work shorter hours than bankers,' he maintained. 'They will never be there when you want them.'

Aircrews, once they get away from Brize Norton, have to function virtually as a self-sufficient unit. They are now issued with American Express cards, but find that angry taxi-drivers expecting cash cannot be calmed down by fanning them with one.

It is not easy, dispensing fuel in flight, nor taking servicemen all round the world. Many a congenial hour in the mess bars at Brize Norton is spent disputing which is *less* easy. I would not be rash enough to judge, preferring to copy that diplomacy which is especially important to the Royal Air Force when it deals with its American allies.

10

The US Connection

'I suppose the most intricate problem in my six years here, alongside the United States Third Air Force, was the bombing of Libya. I am a Ministry of Defence officer set down here in the country at Mildenhall to act not as negotiator but as go-between between the RAF and the United States Air Force and between agencies of the United Kingdom and USA governments. I take messages in both directions, sometimes those they don't want to write down.'

– Senior RAF Liaison Officer with United States Air Force Headquarters in Britain, RAF Mildenhall (a Wing Commander).

'In cases of low flying, the public don't worry too much about the identification of the aeroplane – if it is American or not. They ring us anyway, irrespective of who the aircraft belongs to. It usually happens when an aircraft cuts across a house or a field, not doing anything wrong, but frightening people, especially elderly people. We can't apologize for necessary operational flying, but we often do apologize for the inconvenience caused.'

– RAF Commander and Senior Liaison Officer, RAF Mildenhall (a Squadron Leader).

Question: How many United States Air Force bases are there in Britain? Answer: None.

Then where do those American bombers, fighters and freighters – including the four-engined Galaxy, the largest aeroplane in the Western world – that one sees over East Anglia, and elsewhere, take off and land? Answer: From an example of the durable British capacity for diplomatic juggling with words and circumstances.

Those bases in East Anglia and elsewhere are all *Royal Air Force* stations. The United States Air Force occupies nine major bases in Britain: Mildenhall, headquarters of the Third US Air Force, which is responsible for all US-occupied Air Force bases in Britain; nearby Lakenheath; Bentwaters; Woodbridge; Alconbury; Fairford; Upper Heyford; and Greenham Common and Molesworth, the last two originally chosen to bear Cruise nuclear missiles.

But the 30,000 United States Air Force personnel, and their dependants, work and live on British soil and British bases. RAF officers who work alongside them will tell you, 'We couldn't possibly give things like our land away.' Each of the major bases has a Royal Air Force Commander and Senior Liaison Officer, usually known as the RAF Commander, who is responsible for looking after Ministry of Defence and RAF interests. But, instead of having a few hundred servicemen under him, he will usually have a personal staff of only three or four. He will, however, retain direct command of the British civilian staff employed at the base – perhaps three hundred or so – though it is the Americans who will decide what tasks these men and women will perform.

This British Commanding Officer, who could be forgiven if occasionally he feels like a mouse on an amiable elephant's back, has, in the jargon of the RAF, 'turned over the broad administration of the station to the United States Air Force'. They come as 'lodger units', but take over the responsibility for the maintenance of the station and the operations carried out from it – such as the controversial bombing of Libya in reprisals against terrorists who had been active against the Americans.

If the Americans want to repair a hangar or build a new one, they pay for it. But they pay no ground rent, and there is no lease giving them the right of tenants. The transaction is made on what the RAF calls a Form 2 – the legal document by which the RAF Commander hands over the building to the U S Air Force for administrative purposes. If the U S Air Force no longer requires the buildings, they are then officially 'marched out' again by the RAF Commander, and the property reverts to Ministry of Defence administration.

If the Americans want, for instance, a new barrack block, the United States Air Force first has to secure its own US funding for it. Then it gets together with the RAF to decide whether the barracks should be built. The RAF Commander convenes a Siting Board, over which he presides as chairman. He must be adept in thinking in two sets of national priorities and will consider, along with relevant experts, the '35 per cent design criteria', which give the size and shape of the buildings but do not include the fine detail.

If the design at this early stage gets the thumbs-up, and it nearly always does, the United States Air Force has to jump a hurdle that may be of comfort to British bureaucrats but which rarely proves very high in practice. It must have 'the necessary planning permission' for the building. This may sound formidable to anyone who has tried to erect a conservatory in his back garden; but the planning permission required is the sort that 'any Crown agency' would have to seek. Crown agencies are in fact *exempt* from planning criteria, though it is Ministry of Defence policy to follow them, barring exceptional circumstances.

One RAF officer with experience of working alongside the United States Air Force described the necessary delicate balancing act between American needs and British susceptibilities like this: 'In Army terms, obviously a Centurion tank can't keep up with all the requirements of the Road Traffic Acts, because they are incompatible with its design; but, so far as it is possible, Centurions are moved over the roads as far as possible in accordance with the Road Transport Acts. It is the same with the United States Air Force and planning.'

Strung out as it is between British soil and American

servicemen, the law cannot, in this sphere, afford to be an ass. There are national sensitivities on both sides to be considered. They are best served – it is thought – by a dearth of arbitrarily laid-down rules.

The criminal law, under the Status and Forces Agreement, enacted in the UK by the Visiting Forces Act, is a case in point. There are no specific rules about who would deal with the prosecution of a civil or criminal case in any particular instance. But none of the agreements between the US and the UK, I was told, are operated at arm's length; and each case would be disposed of on its merits, in accordance with the general principles laid down. As a rule of thumb, no more, if an American murders another American on base *or* outside it, he would probably be turned over to the American authorities, though, off base or on base, it might well have been British police who arrested him. There was such a case at Alconbury, where one US airman did murder another, off base, and faced a US court martial at Alconbury, not a British court. If an American murders a British subject, off base or on base, then the presumption is that he would be turned over to the British authorities. But it remains a presumption: nothing is laid down in clear black-and-white.

The atmosphere on and around RAF bases used by the United States Air Force quickly indicates the sort of transnational delicacies that a common language cannot always conceal. It is very tangible at RAF Mildenhall, which has a dual function as the headquarters of the Third Air Force (situated outside the wire and armed guards of the airfield) and the station itself, a main arrival point for United States Air Force personnel, some of them coming over for long tours, others for sixty-day detachments with their own ground crew.

The two hotels almost directly opposite the main entrance to RAF Mildenhall are British, but could easily be mistaken for American ones. The rather more expensive Smoke House Inn, despite its suggestions of English hams being cured, is a modern motel with nine flagpoles, each bearing the flag of a member of NATO. The Bird in Hand is a complex of chalet-type rooms built alongside the original public house, and has only one flagpole, bearing a single Union Jack sandwiched between two Stars and Stripes flags. It is much used by relatives of US Air Force personnel, and conveys to them the notions of England that the Americans like to cling to. When I stayed there, the décor of the restaurant said much about the American view of Britain and the British; absolutely nothing about the Britain of the present, except possibly the street signs for Oxford Street W1 and Baker Street W1, the latter obviously being included not for any modern associations but as the home of Sherlock Holmes. There were railway station signs for Paddington and King's Cross, certainly; but everything else referred to hallowed traditions. There was an old cricket bat on a wall, with the cricket ball stuck on to it with glue and a cricket bag beside it. There was an old advertisement: *Try Tonny's Lollies –They're Good – Only 1d.* There was a sign boasting By Appointment to Tonbridge School. There was a projector bearing the sign Prof. Watkins Magic Lantern Show, a Rolls-Royce hub-cap and a plastic Beefeater. In comparison, the art deco ceiling lights, with their stained-glass strips, seemed almost contemporary. On the outside of the front door of the restaurant, acting as a blackboard on which were chalked the special dishes of the day, was a reproduction pub sign genuflecting to both modern American and former British endeavours: The Empire Strikes Back.

A fine balancing act between two cultures, indeed. It was small wonder that the occupants of the restaurant, British and American, seemed uncertain of what precise roles to play. A mixed group tried hard not to be national stereotypes, the British members of the party gushing expansively over a neat little tight-lipped American lady who looked so *British* in obviously finding it all a bit much. An American man in a Fairisle sweater and a hairpiece was moodily silent while a middle-aged Briton recounted at length details of some of his contacts with the United States Air Force in the Second World War.

It was almost a relief when two women members of the party, one American, the other British, came down to join the group, and were both asked in quick succession how they were that morning.

American lady: 'Fine, just fine!'

British lady: 'Not too bad; mustn't grumble.'

Obviously national characteristics can survive even the enforced proximities of Mildenhall.

Once inside the base itself (even in the residential and administrative part *outside* the wire fence) and it could be America with the solitary exception of the incongruous British red pillar-boxes and telephone kiosks. Near Missouri Road and Indiana Road is the Bob Hope Entertainment Center, which is not only strongly American but strongly evocative of the America of the 1920s and 1930s, since no alcohol is permitted there.

The Third Air Force headquarters building itself is, from the outside, British in its spartan greyish brick. Internally, it has touches of luxury which one would be unlikely to encounter in any British headquarters, save perhaps that of a merchant bank: lots of gold lettering, bevelled glass in the swing doors to the reception desk, brass-framed photographs of former American generals who headed the Third Air Force (all bearing captions engraved on gilt plaques), handsome carpet.

It was rather different in the office of the Senior RAF Liaison Officer, a Wing Commander with a ready smile and cold appraising eyes, both attributes no doubt of service to him in his delicate job. The office was the size of a small bedroom and at a corner of the second floor (i.e. British first floor). It was devoid of decoration except for a large gilt-framed photograph of the Queen, and a large desk-top plaque, decorated with gilt wings and bearing the name of the Senior RAF Liaison Officer – complete with middle initial, American style.

The Senior Liaison Officer, an unpretentious man, said he had previously had a simple card bearing his name. Recently, the Vice-Commander Third Air Force had decided he must have something better, and had presented him with this imposing nameplate as he tore up its cardboard predecessor. I wondered whether this rather summary generosity was somehow symbolic of a relationship in which the British choose to forget their own national interests in an affectionate embrace?

The Senior RAF Liaison Officer, a navigator by trade though in his present job for six years – long enough to know most of the ropes – shook his head. 'You can't be a successful liaison officer unless you secure the trust of *both* sides. Inevitably, because I am a United Kingdom officer, and am employed by the United Kingdom government, I must take national considerations as a top priority. Apart from personal inclination, I must do that in the final result. But I am here to oil the wheels, not to put sand in the works. That doesn't mean you just have to bear the good news all the time. I have to explain decisions, good, bad or indifferent, both ways. The Americans must be able to relate to me as well. I try to make things work. As long as the Americans feel that is the case, they welcome me at their door.'

The door of the US General is indeed only a few feet up the corridor – and the General is a Presidential appointee who deals on behalf of all three US services as negotiator between the UK government and the US military in the UK. The proximity of these officers is valuable at all times, but especially so in moments of crisis.

I asked the Senior RAF Liaison Officer what situation, during his six-year term, had presented the most intricate problems. I was not surprised when he said the bombing of Libya with F111 bombers, normally based at Lakenheath and at RAF Upper Heyford.

Although the Senior RAF Liaison Officer was understandably reticent about the subject of who knew what, and when, he did point out that he was an intermediary, not a negotiator, between the British and the Americans. He had known of the decision when it had already been taken by the Americans with the permission of the British government, according to strict criteria laid down in advance, and on the basis that the F111s were the only way of accurately pinpointing the Libyan targets without causing unnecessary loss of life in the surrounding area.

'It was a question of response to things,' he said. 'I talked to people in touch with Ministers on the one hand, and the General on the other. The General was not told over my head.'

His main function had been to advise the General on possible knock-on effects of the bombing, in relation to the security of the UK bases – what to do and what *not* to do about it. Should more sentries be armed? Were terrorist acts likely at Mildenhall, Lakenheath and other participating bases? Were Libyan students under training at Kidlington likely to drive their aircraft into major installations?

The advice given was that security could be raised and more armed guards used on the bases, but that they should wear standard uniforms and should not be deployed in combat clothes. 'We didn't want them looking like it was World War Three, and they were ready to go into battle.'

Should the commissary or the school of American service children at Ruislip be covered by armed guards? The Americans thought yes; the British were not too keen. But since the school was on Ministry of Defence property the 'appropriate' arming was officially approved and cleared with the Metropolitan Police.

Generally speaking, within the bases, American personnel may carry arms – and the guards at the main gate, huge revolvers in holsters, always look ready for business. Outside the bases, they may not normally carry arms, but, in special circumstances the Chief Constable of a UK Home Office Constabulary could give limited dispensation.

The security aspects following the Libyan bombing would not have been sufficient in themselves, I was told, to justify carriage of arms off base because there was no direct threat posed by way of retaliation. Even if there had been a direct threat, the Chief Constable might well have preferred any armed duties to be carried out by his own men.

One of the questions the Americans asked the Senior RAF Liaison Officer, amongst others, after the Libyan bombing, was whether they should cancel events in which they participated on or off base and close the bases to all but essential business. The typically stiff-upper-lip British advice was to do nothing of the kind, but to carry on with every engagement as normal. So the picnic and the bazaars took place while an overheated media speculated about what Libya would do (*nothing* to US bases in Britain, as it turned out).

Another vexing problem highlighted by the Libyan bombing was the Parliamentary aspect. The Libyan raid was on a Monday. The following morning was Prime Minister's Question Time in the House of Commons. The Prime Minister needed to have for use a clear technical explanation of why British bases had been used for the American attack, rather than a United States Navy aircraft carrier. She was already aware of the argument that only the F111s had the precision navigation equipment to hit the target, and that they were in East Anglia. But more information was fed to 10 Downing Street, via the Ministry of Defence, about the bombing. The Senior RAF Liaison Officer was part of this chain of information.

In such cases, at the same time as British sources are being primed with the facts, the United States Third Air Force General is probably being counselled on how *not* to get involved in any of the issues or public statements. From time to time, the Senior RAF Liaison Officer will remind the headquarters staff that, if they receive letters from members of the British public, or from Members of Parliament, their proper course is to acknowledge them briefly but seek formal response from the Ministry of Defence and UK Ministers. Generals may come and go; the principle seemingly goes on for ever.

Comparatively trivial incidents can often conceal trip wires. Just before I visited Mildenhall, a Woodbridge hotelier had written angrily to the Senior RAF Liaison Officer, complaining that an American officer had booked into his hotel, cancelled at the last moment and not settled the bill in full (£148 50p) as, in English law, he would be compelled to do if the hotel had not been able to find a substitute booking. The American client took the view that he had sacrificed his deposit, and that was the end of the matter. By this time, he was back in the USA, where his legal advisers told him to take no further action. The Woodbridge hotelier complained bitterly that he couldn't sue him.

'Actually he *can* sue him – in the American courts,' said the Senior RAF Liaison Officer, well aware of the delicacy of such situations in which MPs may eventually ask Parliamentary questions. 'It might not be worth it. His complaint is before the Foreign Claims Commission, downstairs in this building. The facts of the matter take quite a bit of explaining. MPs are apt to ask why the United Kingdom government doesn't force the Americans to pay their debts before they go. Fundamentally, it is for the same reason as the Rover motor company doesn't force its employees to meet the legitimate demands of the Electricity Board – they have no powers to do so. People think that, merely because you are in the military, you can force people to pay their debts, or that the military should bear the liability for the private actions of employees. I could wish that the Ministry of Defence would accept this liability, for I would get them to pay off my mortgage! The policy of the RAF is to do no such thing. No government agency would do it.'

By now I was beginning to see why the Senior RAF Liaison Officer had scoffed so violently against the theory that liaison officers to the Americans have to be merely smooth front-men. ('You mean smoothing people down at cocktail parties inside and outside the base? Doesn't happen!')

If there is one area where smoothing-down skills *are* useful it is low flying. There are still about fifty complaints a year coming into RAF Mildenhall as an individual station, plus those arriving at Third Air Force Headquarters, concerning the other United States Air Force flying units in the UK. And this despite the fact that I was told they had been halved in number over the past year. Complaints about the aircraft noise at the airfields or in the airfield circuits had also fallen, largely due to the installation of quieter engines and other means of reducing aircraft noise on the ground.

The Senior RAF Liaison Officer waved an eleven-page report at me as an example of the sort of investigation that had to be carried out in any complaint about aircraft noise, whether from low flying or from the ground. An Independent Broadcasting Authority transmission station in the north of England, used to transmit ITV television programmes, complained that a United States Air Force A10 aircraft (a big one) had flown within a thousand feet of the transmission tower, and as low as four hundred feet above the ground.

There had been interviews with aircrews and with staff officers on the nearest station. But the upshot was that the final report said that no breach of flying regulations, or pilot error, had been revealed.

Did people suspect a whitewash in such cases? The Senior RAF Liaison Officer hoped not. 'I have been flying for thirty-four years, and I would be hard put to it to judge how close an aircraft was to the ground, especially an aircraft as big as an A10 going over hilly ground. I do not think I would know the difference between four hundred and eight hundred feet. There is no percentage in us trying to do a whitewash. We don't spend the sort of money necessary to prepare that report to do a whitewash. Depending on what you find out in the early stages, you investigate deeper and deeper. If there is a flagrant breach of rules, the chap will be taken off flying altogether. I have seen that happen.'

He cited one example. A pilot diverted from the laid-down route, to show off to his relatives on the ground that he was flying an aircraft. Immediately he landed, he was met, grounded permanently and sent back to the USA – in less than the triumph he had hoped to signify to his relatives as he zoomed over their house. The official line, of both the RAF and the USAF, is that multi-million-pound aircraft are not playthings, that being off the ground is inherently dangerous and that nothing must be done to increase the danger.

Both the RAF and the USAF are shrewd enough to know that the fullest possible information is the best public relations approach to local irritations, especially over aircraft noise. It is up to the RAF Commander concerned to advise the Americans when and why it would be best *not* to fly, except of course when absolutely essential. Normally flying does not take place before six in the morning or after ten at night; but there have to be exceptions. One exception at Mildenhall was when, one Saturday night, the RAF asked permission to put a medical flight into the base with two patients from Gibraltar, destined for a Cambridge hospital. There were in fact no complaints that night; had there been, the *reason* for the flight would almost certainly have defused the situation.

Normally local irritation falls upon the individual RAF Commander. The forty-two-year-old Squadron Leader who was RAF Commander at Mildenhall told me that he also did most of the socializing outside the base, from which the Senior RAF Liaison Officer tended to distance himself, confining himself to talking to the Rotary Club if they specifically wanted him to.

The Squadron Leader told me he as very much like the Station Commander of any other RAF station, except that he had the responsibility of liaison with the Americans. He appeared at local council events, and cultivated links with the police and the fire brigade. His wife got involved in Women's Institute activities and community associations, often involving him as her husband. The major US bases each had an Anglo-American Committee, and a community relations adviser, and RAF Commanders were closely involved in Anglo-American activities.

What above movements like CND? Did he take initiatives there?

'Much of the time,' replied the Station Commander drily. 'There are two types of CND, overt and covert. The covert usually takes the form of cutting the perimeter fence, that being their sole act. Sometimes they will come on to the base and spray paint on to a building or an aircraft. Sometimes aircraft are close to the road – though if that aircraft is carrying something sensitive or waiting to go somewhere, there is a guard there all the time.'

Part of the trouble on stations like Mildenhall is that a *Keep Out* notice, which might have been respected in the war and immediate post-war years, is now not an automatic deterrent – on airfields or anywhere else. The conspicuous revolvers of the guards at the main gate have helped persuade CND demonstrators – the overt side of the CND – to stop short there. Four demonstrations a year outside the main gate are average for most USAF-occupied bases in Britain, though Greenham Common has been subject to constant demonstration by an assortment of women who in the 1980s made it a way of life.

For the average RAF Commander at each of the USAF occupied bases, political ideologues are less of a problem than the fine print of the agreements on practical arrangements with the Americans and on the discipline of the civilian British workers.

Financial agreements, dealt with at Third Air Force and the Ministry of Defence, provide that the Americans do not pay rental as tenants of operational bases which are made available for them as part of mutual defence. On the other hand, the USA does pay rental for domestic housing made available by the Ministry of Defence for the use of US personnel.

Domestic buildings are offered to American administration unfurnished, and the rental for accommodation that would normally be paid by British personnel is 'abated' to allow for the fact that the US military do not use British furniture and that the USA accepts total responsibility for the maintenance and refurbishment of the properties. If the housing thus supplied becomes surplus to American requirements, it is returned to the Ministry of Defence for either re-use or disposal, as the Ministry decides.

Fortunately, dealing with the British workforce does not

pose quite so many conundrums. One of them is proper dress. Despite their reputation for informality, the Americans have probably been overtaken by the British in matters of informal dress. At the air terminal, through which arriving United States Air Force personnel have to pass as if they were at a civilian airport with a customs desk, it is an American-made rule that all staff wear white shirt, grey trousers and blue blazer. In case of persistent refusal to adopt this uniform, there is a written reprimand which will tell against a person's promotion prospects.

If a British worker is unpunctual, it will be the RAF Commander who will deal with him, not the Americans who have actually diagnosed his unpunctuality. If he is overfond of a pint or three at lunchtime, it is the RAF Commander who will ultimately bar him from the base, not the Americans. The RAF Commander is responsible for all security questions concerning Ministry of Defence civilian staff, a task which may well seem more real to him than the fine print of other agreements.

For one only has to tour the airfield itself to remind oneself of what the reconciling of little, big and potentially big national differences *is all about* on bases like Mildenhall. It is about having a lot of supply and fire power on British soil, as a vital part of the defence of Britain and Europe. When one bears in mind the enormous war machines massed by the Warsaw Pact countries, nuclear and chemical as well as conventional, the sight of an enormous Galaxy trooper-freighter aircraft – so much larger than anything Britain has got, with a tailplane higher than many a block of flats – somehow redresses the impression of an unequal balance.

The SR71 Blackbird, a pencil-thin black tube of an aircraft with two enormous jet engines, capable of *three* times the speed of sound, easily out-performing the European civil Concorde, is another visible equalizer. So is the EC135 airborne control aircraft, which (unlike many RAF aircraft) can both refuel other aircraft *and* be refuelled itself.

As I took my leave of this potent if unlovely fire power,

the Four Seasons Shopettes, the hand driers in the washrooms labelled 'Fragile', the Ye Olde English décor fighting with American-style neon lighting – indeed *all* the manifestations of the time-honoured bi-national mix – did not seem quite so risible. They had a point. Self-interest on both sides, no doubt; but a self-interest that has helped to keep peace in Europe for nearly half a century.

The head must endorse what the heart may sometimes rebel against as wounding to national self-importance. It is possibly that reasoning that helps the carefully recruited and trained Royal Air Force man and woman to work diligently in cementing Anglo-American relations whenever the occasion arises.

PART III

WAYS OF ADVANCEMENT

11

All Classes Welcome

'An RAF officer candidate will not fail simply because he has the "wrong" social or regional accent. But you have to remember this – his Liverpool accent doesn't matter when he is at home in Liverpool, but when he comes into the RAF, he must be easily understood by people from all regions. In accents, an officer has to come towards the middle ground; we must all, in a sense, be cosmopolitan.'

– Commandant, RAF Officers and Aircrew Selection Centre, Biggin Hill (an Air Commodore).

'I've just seen a thirty-two-year-old NCO, who wanted to become an officer, being interviewed. He kept correcting himself. "No, sir, I tell a lie . . ." He said he was interested in exerting intellectual leadership, but he hadn't *done* anything about it. He was eager to point out that he read a right-wing newspaper. I think that finished him. He *wasn't* what they were looking for.'

– Guest observer at Officers and Aircrew Selection Centre, Biggin Hill (a master at one of the top twenty public schools).

As the Squadron Leader drily put it, Auckland was not exactly a straightforward candidate. 'No, he isn't,' agreed the Wing Commander, with carefully expressionless fairness.

Auckland, according to some of his character references, provided to objectify his ambition to become a pilot in the

RAF, was nice, trustworthy, kind, pleasant and quickly able to acquire new skills. He had indeed proved that last compliment. In a sense that was the trouble. In the course of his twenty-five-year-old life, either in Britain or in the country of his birth, in the old Commonwealth, he had been employed as a furniture repairer; a child care worker; a financial consultant on portfolio investment (for three weeks); a youth worker; and a law-enforcement officer.

Auckland was a tall young man with a rather creased grey-suit, well-polished black shoes, socks slightly out of the vertical and a self-deprecating laugh. He was not, at first sight, ideal material for a disciplined fighting service; not the typical product I had rather assumed the Officers and Aircrew Selection Centre at Biggin Hill, in Kent, would be looking for.

But practically the first moment I – and Auckland – had walked past the twin Spitfires beside the main gates of the Battle of Britain airfield that is now used for a rather different peacetime purpose, the Commandant (a grey-haired Air Commodore who would have sat as easily behind the desk of a headmaster or an industrial executive as he did behind his own) had assured me that the RAF's policy, in looking for its future officer material, was to 'cast its net as wide as possible'.

Predictable words when other services have sometimes been criticized for being too élitist in their search for officers? Auckland, and the five other young men who formed one of the syndicates of candidates I saw through the three-day selection process, was proof of how wide the net can be cast. Out of 8,000 would-be candidates, some 5,000 candidates a year – men and women – pass through Biggin Hill (most, like this particular six, never having seen the inside of a public school) in order that the RAF can make its final selection of about 1,200.

It is tough going for the officers who make up the selection boards, as well as for Auckland and his fellow candidates; three days that must get as much adrenalin pumping at Biggin Hill as in the days of the war, when Hurricane and Spitfire pilots took off from there to destroy the Luftwaffe's hopes of gaining control first of Britain's skies and then of Britain itself.

There is indeed plenty in the history of Biggin Hill to make it receptive to any necessary expediency, however unconventional. On the first day of the evacuation from Dunkirk of the British Expeditionary Force, in May 1940, three fighter squadrons were based on the station. All three were in action. In the next nine days, fighting to provide air cover for the British troops, the squadrons shot down thirty-six enemy aircraft for certain at the cost of fourteen Hurricanes and Spitfires. From 12 August 1940, when the Germans started their large-scale attack on fighter airfields that signalled the start of the Battle of Britain, Biggin Hill was heavily bombed; many sections had to be moved out to other areas so that the station could remain operational. Only when the enemy attacks were switched to London itself, in September, was there an easing off of the punishing life-style at Biggin Hill. By the end of 1940 – and of the Battle of Britain – the Biggin Hill squadrons had destroyed six hundred aircraft. During the war as a whole, more than thirty squadrons flew from the station, shooting down 1,400 enemy aircraft and earning two hundred decorations for gallantry and outstanding feats in action.

There are permanent reminders of that fraught and gallant history on display at Biggin Hill today. In the officers' mess, across the road from the station itself, there stands, at the foot of the stately oak staircase, an enormous framed and glazed panel of Nottingham Lace. It is twenty feet high by six or seven feet wide, and rather difficult to miss. It was presented in 1947 by the makers to the Borough of Beckenham, as one of the towns in the forefront of the Battle of Britain. On 1 April 1965, says a brass plate, Beckenham became part of the London Borough of Bromley, and on 7 April 1970, the panel was handed over for 'safe keeping', and as an expression of gratitude, to Biggin Hill. The panel shows St Paul's Cathedral, British and German aircraft and parachuting pilots, and includes the famous phrase of Winston Churchill's, 'Never was so much owed by so many to so few.'

It is perhaps fortunate that the officers' mess is out of bounds for the candidates, who are expected to socialize as a group across the road in the mixed-sex candidates' 'lounge': they are not expected to think *completely* in RAF terminology until they are actually in it. The sight of that huge lace panel could be intimidating, speaking eloquently, as it does, of Biggin Hill's history.

Today the awareness of that proud history must make the more imaginative and sensitive of Biggin Hill candidates, even without a sight of the panel, at once more stimulated to do well and nervous that they may not manage it. Only a few candidates remove themselves from the battle almost before it has started – usually those suffering from drug abuse or those trying to test the system on behalf of CND. Officers at Biggin Hill say that 'character defects or holding views incompatible with military service' are the two things most likely to betray a candidate as worthy of an 'unsuitable' rating.

Anyone who is deemed better than 'unsuitable' (and there are seven progressively more flattering grades) has a chance of selection. Certain minimum levels of ability and suitability in objective testing must be achieved, but above that there are no absolute pass/fail criteria. The Commandant selects the best of those available in competition.

The pleasant and versatile (in civilian terms) Auckland, keen to be a pilot after the two counselling interviews and the sight of an eleven-minute slide presentation given in advance to all candidates, began his three-day boarding routine on a Sunday with a medical examination and with aptitude tests. It was to emerge later that he had fluffed both.

True, Auckland did not fluff the pilot's aptitude tests as badly as I did when I took the same tests, for the sake of comparison, directly afterwards: the standards demanded of people who are to fly £20 million worth of aeroplane are understandably high. The tests, now highly computerized, are held in a separate building which still sports (for nostalgia only) the mechanical cockpit mock-ups and other clumsy mechanical testing devices of former years.

'It was in the mid-1940s that aptitude tests were first developed,' said the Squadron Leader in charge of the tests. 'The result was a 50 per cent drop in the training failure rate towards the end of the war. Our task here is exactly the same as it was in the war; but nowadays our restraints are financial rather than operational. Of the thirty-eight expected here for the board, two haven't arrived and six will drop out from the aptitude test – that is about par for the course.'

Candidates used to have pencil-and-paper aptitude tests on such matters as mechanical comprehension and co-ordination. The *principles* are still applied, but in batteries of tests by a virtually completely computerized system. 'We can pick out five different aspects of ability with one test,' said the Squadron Leader. 'The old mechanical tests were on machines that could vary by 20 per cent. We have forty-four computers here now and we have saved 40 per cent of the manpower required to run aptitude testing. If we can save one man from going through to fly Tornados who should not have gone through, then we can save up to £3 million in training costs.' Crashed aircraft too are expensive.

To misquote the Duke of Wellington: I don't know what the aptitude tests did to Auckland, but, by God, they frightened me.

They last four and a half hours, and demand a co-ordination of mind and limb which would leave the average civilian far behind. In one test for pilot aptitude, I sat in a large room given over completely to video screens, each one in a cubicle. With an electronic joystick in my hand, I had to knock out as many as possible enemy planes or missiles in a simulated battle. The very minimum score expected of a pilot is 110. Auckland got ninety-one, which was rather more respectable than my eighty, but not good *enough*. A

Ministry of Defence psychologist who was also observing the boarding as a guest, like me, managed 113. But another guest, a master at one of the top twenty public schools, managed only seventy-six. This somewhat comforted me, but couldn't comfort Auckland, as he didn't get to hear about it.

All Auckland heard before the next step the following day – the personal in-depth interview – was that he had failed his aptitude test as a pilot, *despite* holding a valid civil pilot's licence for which he had done the required flying hours.

Did Auckland want to stay in the boarding process for his second choice, that of navigator? He did. He was called for his individual interview, held in one of twelve rooms specially designed for effective and efficient interviewing.

The interview lasted for forty-five minutes, a substantial time to fill with carefully studied pretence. The Squadron Leader kicked off with questions about Auckland's background and educational qualifications, the first one being the sort of easy question designed to put the nervous candidate at his ease.

'You were outside Britain when you received your notice to attend here. How did you get back here?' asked the Squadron Leader.

'I flew.'

'What was the journey like?'

'Nice trip. Chance to read and relax.'

Already the deceptively easy questions were telling the Squadron Leader and the Wing Commander, whose turn was to come later, something about the potential officer before them. The second question, about how the British railway system compared with the one in Auckland's country of birth, revealed some more: 'I find it efficient. Over there, there are such long distances to travel that they concentrate more on flying than railways.'

Yes, Auckland obviously had eyes for what was happening around him as well as a desire to please.

'What is the accommodation like here?' asked the Squadron Leader.

'Comfortable. I am enjoying it.'

'Was it what you expected?'

'Yes. I was not sure what to expect, but when talking to a relative, he said I would be in rooms four feet by three – so there we are!'

Auckland had a sense of humour that might be expected to stand him in good stead in an uncomfortable situation. And the questioning, apparently simple, went on until it had covered the whole of the background of Auckland (whose real name was not Auckland, any more than the names of his fellow syndicate members I will mention will be their own; furthermore, as another barrier to identification of individuals, some of the facts will be changed, though the total impression will be a true one).

The final question from the Squadron Leader was terse. 'Drugs. Ever been tempted to try it yourself?'

'No,' was the firm response. 'I have not been near it, nor tried it. People who are breaking the law, they are destroying something that is invaluable, the body, and I have no time for it.'

Even the factual answer said something more about Auckland than the mere facts: here was a young man with a sense of order and, perhaps more important, a religious sense of what was important and priceless about human life.

The Wing Commander, a quiet man with battered spectacles on a face that could have passed easily in the Civil Service or the senior common room, took over, again with an easy question.

'How many hours' flying have you had?'

That established, the second question was more pointed. 'Unfortunately, you failed your pilot aptitude tests. How is that going to affect your thoughts about the RAF?'

The answer came with a brave assurance that was to seem ironical at a later stage of the tests, when there would be other disappointments for Auckland. 'I was really disappointed, but I am quite happy with the chance to be a navigator, because I will still be in the plane.'

'You are sure about that?'

'Over here pilot and navigator are like an equal team, but over there, where I was born, the pilot is the hero and the navigator is the "Joe-carry-your-bags" type of thing. The RAF is just something I have always thought about – wow! I want to be a pilot in it. The machines they fly! How the Luftwaffe was sent away, that sort of thing. But to be in it at all . . .'

'What sort of aircraft would you like to fly, as a navigator?'

'I think I would like to fly a Tornado.'

'What do we use Tornados for?'

Auckland got the answer wrong – and indeed was often, in the boarding process, seen to have the right spirit rather than the right answer. But officers pointed out to me that knowledge, unlike deep human attributes, could be fairly straightforwardly taught.

Other questions followed, designed to test Auckland's awareness of what the RAF was and did. Then suddenly: 'How many hours a month would you expect to fly, as a Tornado navigator?'

'Thirty hours.'

'And what else is a keen navigator doing with his time? Why do you think a Commanding Officer gives people extraneous tasks to do?'

This was a delicate question: in both the RAF and the Navy, two heavily technological, specialized services, officers and men now have to 'double up' on their set tasks, doing an often mundane second round of duties as well as their prime function. It is a phenomenon which sometimes breeds resentment in even the most diligent.

Auckland made short work of the question. 'You have to prove your worth. You cannot just sit around when your flying is over. You have got to be prepared to do other things, administrative duties, as well.'

The Wing Commander moved swiftly on to questions designed to test the candidate's knowledge of, and attitudes to, the modern military world. The Falklands eventually arrived on the agenda. Why were we in the South Atlantic?

'Because the Argentinians figured that what they call the Malvinas were their islands, and they took over, so Britain sent out a task force.'

'Send out a gunboat and raze it to the ground; British diplomacy, eh?'

Auckland acquitted himself well on this example of the art of the agent provocateur, thus proving that he could do the same if chatted up by some friendly Russian spy in a pub. He laughed. He *said* nothing.

'Why do Eastern bloc people create a threat in Germany?'

'Because they feel their way is the way to be and, if they want something, they just go and take it. I mean the Soviets, the Warsaw Pact countries.'

Some questions on international affairs followed. Auckland made no pretence, in his answers, of being a conventional dyed-in-the-wool militarist of the type so beloved of the satirists. He thought some judgments were based on 'the American propaganda machine'. He thought the American journalist who had just been seized in Moscow as a spy probably *was* a spy, though the Americans denied it – but he still thought the American President was right to turf a number of Russian 'diplomats' out of the United Nations Headquarters in New York.

In short, Auckland demonstrated that he was not a naïve nor violently partisan person but, when the chips were down, would prefer his own side to an enemy's (a realism by no means to be found among all liberally minded people).

After being asked if *he* had any questions (he hadn't) Auckland was politely shown the door. Then the Squadron Leader and the Wing Commander discussed what they had heard and seen in minute detail.

'Imagination quite good,' said the Squadron Leader.

'He had gone carefully into things, and thought them out,' said the Wing Commander. 'But his awareness of what was going on was shallow.'

'Yes, I gave him headline-quality current affairs. But he has had bigger areas to cover, given his background.'

'Quite a nice lad, really,' said the Wing Commander. 'There is not much wrong with him. I think we are of the same view? Personality quite good, plenty of determination there, very well rounded and balanced.'

'Presentable, versatile. Weaknesses? Little group involvement in his background.'

'But there is a preponderance of good things here,' insisted the Wing Commander. 'I like him. There is a nice warmth about him, which comes over even in forty-five minutes. A bloke you are grateful to have around you in the mess.'

Alas, Auckland's luck was not to last. By the beginning of the afternoon of the second day's boarding, I was told that Auckland had failed his medical for navigator – another blow after the failure of his aptitude tests for pilot – but had decided to stay on to be considered for air traffic controller.

This meant he went forward for his next tests under a different Wing Commander, who could bring a fresher view to assessing Auckland in action. The tests started off in the Syndicate Room, where Auckland was joined by the other five members of his syndicate – who were all about to go over to an enormous hangar in which were arranged various forms of puzzle courses. These involved the transportation of objects over distances, helped only by rudimentary equipment, ingenuity and co-operation.

All the young men wore, on their backs, large numbers, from one to six, allocated in alphabetical order, and were addressed throughout the exercises only by their numbers. I shall use fictitious names rather than numbers, for the sake of easier comprehensibility.

There was Brooks, a Londoner of twenty-five, living with his parents and working for a cleaning firm. There was Ladd, the youngest of the group at nineteen, a salesman still living in the parental home, a tall, thin, pale young man. There was Mann, aged twenty-six, who had been earning £20,000 a year with a financial firm until it merged with another, and he was made redundant. There was

North, product of a north of England polytechnic, and now a clerk with a nationalized organization: his accent was inescapable. The fifth was Auckland, and the sixth was Stark, a twenty-three-year-old engineer.

Although referred to only by number, the six quickly established themselves as personalities in their first, leaderless, talk in the hangar. Being without an appointed leader, as the Wing Commander pointed out to me, is difficult; even if you have a *bad* leader, at least you know who is to sift and synthesize ideas and pronounce the decisions.

I quickly saw what he meant. Left to themselves, to see who made the first concrete moves, it was the oldest man in the group, Mann, who first picked up a plank to see where it would stretch and where it wouldn't. Auckland was standing back from the action, making suggestions to the others; North was correcting these suggestions (often mistakenly); Ladd's immaturity was plainly telling against him, as he was virtually shoved aside and ignored; Stark came in with some incisive suggestions, and stood by them when criticized.

This sort of pattern was to continue throughout the exercises, with the proviso that the quiet Auckland maintained his calm under pressure whereas the immature Ladd once lost his temper and shouted, 'Let's get someone over to the other side first, not sit talking about it!' – without indicating how this might be done.

Time up. The Wing Commander made to them the point he had made to me about leadership: 'You see, gentlemen, even though a leader may not be the most effective man in the world, at least you have a man to assess and filter ideas.'

To me he said: 'The only one who gave me real worries was Ladd. Every time he tried to do something, he was overwhelmed. The two strongest were Mann and Stark.'

As a civilian I had been forced to see the same thing, though in fact – to me – Mann and Stark were possibly the least ingratiating personalities of the six; the least likely to charm a civilian professional circle. I had, much earlier, observed the same sort of thing when researching my corresponding books on the British Army, *Soldiering On*, and on the Royal Navy, *Ruling the Waves*: civilian ideas of what 'brightness' consists of have to be revised in the armed services. Possibly they have to be revised rather less in the Royal Air Force than the other two services – the RAF has perhaps the closest relationship with purely civilian values – but they *do* have to be revised. It is one thing to have ideas, quite another to be able to act on them, and influence *others* to act on them, while under mental and physical pressure.

The discussion exercise that followed proved the point clearly. Each candidate was free to offer opinions on a variety of propositions of the 'Space exploration is a waste of money' variety. It was the youthful Ladd, who had been swamped in the leaderless exercises in the hangar, who spoke up first and who sustained his points throughout with the greatest sense and clarity. Indeed the individual performances in the discussion exercise were in almost exact inverse proportion to those in the leaderless task exercise.

Ladd, bottom of the six with 10 per cent in the leaderless task, led the discussion field with 60 per cent. The second weakest in the leaderless task, Brooks, was the second strongest in the discussion. Only the possibly discouraged Auckland did indifferently in *both* action and discussion. The general picture was a clear one: thought and action can be two very different things, and an RAF officer needs proficiency in both.

The planning exercise was devised to test the ability to plan a practical course under the pressure of an emergency. Obviously the RAF is not keen to have its tests described in too specific detail, because candidates would then know exactly what to expect; but the exercise faced by the six-man syndicate I was watching concerned a notional situation in which the syndicate were executives in an overseas office of a business who were facing 'national unrest inspired by Communist-led agitators', and had to decide what to do about it with the stated resources available to them.

In the public discussion between the candidates, of the possibilities, all the syndicate weighed in with ideas except poor Ladd, who was doodling uneasily on his jotting pad, reduced to asking questions of the others. Questioned by the Wing Commander, to see if he had grasped the essentials of the problem, Ladd got the name of the nearest town wrong. Auckland made a suggestion about what to do with the restless native workers which was instantaneous but ill-advised. North forgot whether the women and children he was supposed to be saving were coming or going. Mann suggested some bold courses, not all of them practical. Brooks faded away. Stark made little impression.

While the syndicate members were individually producing a written version of the finally agreed plan in rounded English, the Squadron Leader and the Wing Commander discussed the candidates.

Of Brooks, the Squadron Leader said he showed some degree of promise, but he 'thrashed into things without sorting out his priorities'. He made statements of the obvious and repeated himself; and, left to himself, was unlikely to come up with anything very solid. But he did seem to work calmly under pressure.

'He tends to be mouth before brain,' said the Wing Commander.

The Squadron Leader suggested a low score for Ladd, because he hadn't contributed anything. 'When he piped up, he was put down.'

The Wing Commander broadly agreed. 'Some points, I think, for his understanding of the questions, but he didn't contribute anything.'

Of the ex-£20,000-a-year Mann, the Squadron Leader said: 'Happiest in support – looking for a boss. I like his rational approach – he doesn't jump in with both feet. But he's a bit casual. Deferred to others readily. A sense of humour. Okay under pressure. I expected more from him in the overall plan.'

'Short on drive, assertion and projection,' said the Wing Commander. 'His contribution was rather sporadic. We will mark him down slightly for that.'

Next, the north countryman North. 'He saw some of the finer points, and he had a willing sense of purpose, but was lacking in the ability to sustain his influence,' said the Squadron Leader. 'Very modest, and just can't get himself across. They didn't pay much attention to him.'

The Wing Commander thought that about right. 'I think, in involvement and effort, he is not up to standard.'

The Squadron Leader was 'quite surprised' about Auckland's resourceful ideas. 'They allowed him to be the co-ordinator. Good overall comprehension; but Stark came in and took over. Fades under pressure. Only fair eye contact. I give him his due for initial involvement, with a high average score.'

'I will agree,' said the Wing Commander. 'He came over quite well, and kept on going, although some of his early influence faded away.'

Finally Stark. The Squadron Leader thought he took initiatives and showed imagination. But he deferred to others. 'Strong eye contact,' he recalled. 'He was looking *us* down. A realistic sense of priorities. I have him down as the top scorer.'

The Wing Commander agreed; and the two officers went on to finalize what are called the 'overnight' marks of the candidates. These are the marks that can be overturned in the light of what happens on the final day of the boarding process, when the candidates have to do their command task, each one taking it in turn to be leader. On their tests so far, Brooks stood at average, Ladd at low average, Mann at good average, North and Auckland at good average, and the bleak-looking but incisive Stark at above average, easily leading the field overall, as he had led the field in the planning exercise, compared with the runner-up Auckland.

The two officers adjourned to the office of the board president, a senior officer who agreed all the results, in this case, after hearing the justification put forward by his subordinates. The president allowed the occasional flash of humour about the candidates, always balanced by a kind awareness that all were under strain.

The Wing Commander pointed out that one of the candidates had had a child six months after he got married.

'But,' chipped in another board member, 'he still hasn't found out what caused it!'

The quip did express, in a pointed way, a certain lack of awareness in that candidate's make-up.

Only one chore remained before the candidates could return to the candidates' lounge for the evening where they could mix easily with male and female candidates (and possibly discuss such subjects as what causes babies). This chore was the signing of simple forms to signify that each of the candidates had read the leaflet saying what would be done by way of security vetting. Signing, pointed out the Wing Commander, meant *only* that they had read the leaflet, *not* – at this stage – that they were agreeing to its provisions.

'All right, gentlemen,' said the Wing Commander genially, 'I hope to see you, bright-eyed and bushy-tailed, tomorrow morning.'

They were possibly more bright-eyed and bushy-tailed than I was when the individual problem sessions started at just after half-past seven the following morning. Each of the six candidates in turn was put the same problem, for which he had to suggest a solution. The problem concerned getting people back from a complicated holiday location, ahead of possible industrial disruption. The candidates provided fewer surprises now: where they would succeed, and where they would fail, was more predictable in the light of the impression of them already gained. At least, this was so as far as I was concerned, though the board officers strove to keep open minds and judge each candidate meticulously on eight different qualities. The qualities were initial confidence; recall of facts without having to consult notes; comprehension; perception; accuracy; receptiveness; mental agility; and reaction to pressure. Particular weight, I was told, was given to comprehension, perception, mental agility and reaction to pressure.

The perky but sometimes superficial Brooks prefaced every answer he gave to the board on how he would solve the problem with, 'Well . . .' or 'Right!' which in the end became as agonizing as having teeth filled.

Ladd, the youngest man, who had so often been overwhelmed in group action, came up with good solutions, put them forward in a quiet but firm voice and proved mentally agile under pressure. This, I was to learn, did not please the board as much as it pleased me: to them, it indicated that Ladd was a *solitary* thinker, the sort of man who might be happier working on his own on problems, not with other men. But Ladd easily led the field in this exercise, quite unlike Mann, who had been regarded as much more promising material, but suggested a course of action likely to be decidedly risky. North provided little of the sought-after eye contact, and had to admit that he had not been able to finish off his calculations in the time available.

The pleasant Auckland became confused and rambling under pressure from the board, and had to alter all his vital calculations under questioning.

Finally Stark, who had done so well in action, did badly when faced with an intellectual and practical problem on his own. One of his suggested plans could have entailed a seventy-mile walk for someone.

One of the officers said drily to me: 'In 1940, during the Battle of Britain, they would have been better, because they *taught* mental agility then – they don't now.'

But the officers were more concerned about deciding the 'batting order' of the candidates in the hangar – for their exercises in commanding the others – than with analysing the poor results sometimes obtained in the individual problem section.

The intense and unsmiling Stark took command first and shrewdly assigned to the also-strong Mann the most difficult tasks. He used polite expressions like, 'Would you care to . . .?' rather than abrupt orders, and at one point asked advice of the watching Squadron Leader, only to be told shortly: '*You* are in charge.' Before the end of the test, with the usual planks, ropes and petrol drums, Stark was reduced to asking '*Anyone* got any suggestions?'

Perky as ever, Brooks gave a long set of instructions to the others in a long and rambling speech, which baffled the board officers as much as it baffled the other men, who couldn't remember a tenth of it, and could not have implemented the 10 per cent they could understand. Falling bodies and general disaster greeted Brooks's efforts.

Poor immature Ladd fared even worse. He ended his briefing of the other men, detailing his scheme, with, 'And after that I will have to ask for your suggestions as to how to proceed further.' At one point Ladd ordered Mann to come back to the starting line under the penalty rules, and Mann refused – and got away with it.

'This one is desperately poor,' the Wing Commander whispered to me. 'They *all* are!'

The northerner, North, gave a clear, short briefing and kept control of the other men, but got them all into a terrible muddle. 'Well, he was *trying* all the time,' said the Wing Commander fairly.

Auckland had to count off the special rules of the exercise on his fingers, but deliberately provoked a laugh about it which eased the tension. At one stage he said, 'Sorry about that – my fault.' He was quiet, but usually incisive.

Finally Mann was in charge, the oldest member of the syndicate. He gave his briefing clearly and good-humouredly, and managed to keep control of his men, even when they all fell over and had to start again.

In the individual interviews with the board which followed, each man was asked how he had regarded the whole business. Brooks: 'Tremendous fun.' Ladd: 'I think my visit here hasn't been as good as it could have been. I don't think I have performed in any way as well as I could do. If I apply again, will my performance this time be taken notice of?'

No, the Wing Commander assured him. He would be judged entirely on his new performance.

Mann said he had enjoyed it all, and thought he had learned a lot about leading people. North found it all 'quite stimulating' and thought he could handle more pressure than he had anticipated. Auckland admitted that sometimes he could have used more time. The intense Stark claimed he had now learned how not to take things so seriously.

The board shut themselves away to assess the men for the whole of the hangar exercises. The markings at this stage have a great deal of subtlety. Not only can candidates be awarded straightforward ratings from nought to seven; they can be given borderline markings. A three-two means basically a three, with a question mark over it; a one-two means a one with the possibility that the man is being underrated.

Brooks was given an unambiguous low average rating which presaged rejection. This meant, 'Discourage him.' Ladd got a low grade with a recommendation to wait two years before applying again. Mann was given a rating which meant, 'Yes – just.' North was given a lower grade, which meant, 'No, but try again after a year.'

Auckland fared better than most – because at this stage, *not* before, the Squadron Leader and the Wing Commander had access to information on the young man's background. He was initially given an average grade, but this was altered to high average when his interest in private flying and other facts were taken in account. 'He *is* a risk. But he is possibly worth the risk of having a look at,' said the Wing Commander. 'Yes. Just!'

Stark was found to be disappointing in command though good in support and, moreover, he was overweight. But he held up well under pressure, and also got an average grade modified, after his background was taken into account, to a high average. 'I think he is worth a go. Yes. Just worth the risk,' said the Wing Commander.

All these recommendations were put to the board president, who agreed them for passing up to the Commandant, who can either adjudicate on differences of opinion himself, or hold a selection committee to resolve them.

That is the system in action. I was especially impressed by the way the officers taking the candidates through their hangar and theoretical exercises knew nothing about their background until they had already reached not only a view, but also a definite marking. It is arguable that in this sense they can have more unprejudiced objectivity than comparable officers in the British Army and Royal Navy.

All the same, I had a few questions for the Commandant before I left his ex-Battle of Britain domain now adapted to the needs of the times.

No doubt the pilot aptitude tests are a marvel of scientific invention and computerization; but is it fair that a man can be ruled out totally and finally through these tests when the RAF takes the view that many defects can be remedied by training? Officers had assured me that any man who failed the aptitude tests was simply not the sort of person who should be at the controls of a service aeroplane: Auckland might well be capable of holding a civil pilot's licence without being suitable for the RAF's specialized tests. The Commandant bore this out.

'We have expanded the process to allow a whole lot to be proven by clinical examination,' said the Commandant. 'We have a scientific assessment of exactly the sort of person he is, and what sort of job he will fit. We are great ones for the task analysis approach, led by our scientists, by which we continuously examine exactly what such-and-such a person in the service has to do. If there is a new form of technology, then we re-examine the requirements that must be found in the person who is to operate it. If a different skill is required, then a different aptitude must be identified.'

Women as well as men are tested at Biggin Hill. 'And they are tested in essentially the same way,' said the Commandant, anticipating my question. 'We give them lighter planks and lighter barrels, perhaps; but the nature of the exercises is the same, and the things we expect them to do, and the qualities we look for, are the same.'

Both the Army and the Navy will admit that, though

they hope to recruit as many good officers as possible from all sorts of school, experience has convinced them that they will in practice get the most sure result from public boarding schools. The Commandant of Biggin Hill assured me that the Officers and Aircrew Selection Centre would look at candidates who might be, at a superficial glance, the most unlikely material, not at all the most obviously self-confident product of the most self-confident schools.

'In order to find sufficient numbers from which to make our selections of people with the particular qualities we need, we need plenty of applicants, because we have a sizeable number of vacancies to fill,' said the Commandant. 'We will not fail people because they fail in one thing. We will *pass* people because they have adequate strengths to see them through training. Of course, they may have weaknesses in a particular area. There are a whole lot of things that go to make up an RAF officer, especially if he is going to be an expert, like a pilot or an engineer. You may say at first, "He doesn't look like an officer to me." So he might not. But just wait until he has been through training, and is out at the other end.'

The consideration of the results of the six candidates I had seen was just as thorough and time-consuming. With Mann and Stark there was no problem. The Commandant selected both to be RAF pilots. Brooks was found to have character defects which the RAF would have found difficulty in swallowing; he failed even as a potential supply officer. He was free to apply again but, in view of his comparatively high age, was not thought likely to get through. Ladd also failed, but, as he was only nineteen, he was told that the RAF would be quite happy to see him again later.

Only North and Auckland had to be considered in depth by a session selection committee chaired by the Commandant. In the end, Auckland was very favoured for air traffic control, though he had come to Biggin Hill hoping to be a pilot. North was felt to have more question marks over him, but in the end got through – but as a NCO air electronics operator.

'Four out of six passed is above average,' rejoiced the Commandant. 'But I suppose the fact that two were passed for air service and two for ground service makes it about average. I pick up some of my ground successes from people who come for air boards. I keep reminding the RAF that it must keep the pilot flagship sailing. It is from the people who are attracted to us with the idea of being pilots that we pick up so many people for ground service, by the time they have accepted the realities of the test results.'

All the selections had to be notified as recommendations to the Ministry of Defence, but as 99.99 per cent of the recommendations are accepted, Mann, Stark and Auckland were set on their path to RAF officer training, and North on his path to NCO training.

The Commandant was especially interested in what would happen to the northerner, North, when he had been in the RAF for a little while. 'In his case, his limitations are a question of his environment,' he conjectured. 'He reflects the environment in which he has lived all his life – very narrow. If one can get him out of there, then I think we can harden him up a bit. It is now a question of training. He has a lot of qualities for us.'

Yes, I thought, and it might have been even *less* likely that North, with his northern accent and 'corn-cob' philosophies, would have become an officer in the British Army or Royal Navy; and perhaps less likely *still* that, one day in the future, he might be able to switch to an officer's cap.

As I walked past those two watchful Spitfires by the main gates of Biggin Hill for the last time, I cudgelled my brains to remember what had signalled to me most effectively, in the course of watching RAF officer boarding procedures, that the RAF had subtly different requirements, and a subtly different philosophy, from the other armed services – though discipline is, in its own way, just as relentless.

It was when one of the boarding officers said of a particular candidate: 'You know, he's a bit of a wally.' In some quarters of the British Army, this might have meant no more than: 'He's got his glasses on a bit askew.' In some quarters of the Royal Navy it could have meant: 'He hasn't taken the trouble to know beforehand how we do things in the Royal Navy.'

I asked the boarding officer at the Officers and Aircrew Selection Centre at Biggin Hill what *he* had meant. 'I meant that he's a chap who *can't work things out*,' the boarding officer said immediately.

That said quite a lot about the RAF's requirements and its attitude to life. An RAF officer needs to be a thinking and effective person in civilian terms – only *more* so.

12

The Real Classes (1): The Officers

'Baby pilots are tolerated. They are childish, with more group awareness of *themselves*. For pilots that may be a good thing, but I do not think it is good for the Air Force. Personally, I think everyone has their part to play, and it is dangerous for anyone to hold themselves apart.'

– Engineer officer student on engineering management training at Royal Air Force College, Cranwell.

'The Warsaw Pact air forces fly fewer annual hours than the RAF, although they are catching up. I would say to anyone who thinks the RAF can fight the Russians alone: those days are over. We have got to realize that the Americans must remain militarily strong: otherwise NATO will be weak, despite our contribution of highly skilled pilots.'

– Head of RAF Flying Training School at Cranwell (a Group Captain).

The would-be Royal Air Force pilots are known as 'baby budgies' by the ground crew and other recruits, partly in retaliation. It is part of the competitive 'us and them' syndrome which emerges from the earliest days of RAF basic training, a social division which works less according to rank or family background than by professional type.

An engineer, officer or airman, tends to be greeted (or more rarely dismissed) as a 'bluntie'; a supply man, officer or airman, as a 'stacker'. Only officers can be pilots; and the more leisurely and thoughtful administrative types refer to them as 'fast jetters' – the type of people, in other words, who are likely to have their hair-trigger minds sorely tried by lesser mortals.

This applies, I was to discover, to the officers training in the flat windswept countryside of Lincolnshire near Lincoln itself – at RAF College Cranwell, in conditions of some grandeur – *and* to the airmen doing *their* training at Swinderby RAF School of Recruit Training near Newark, the rather more windswept and less prepossessing site of a less attractive group of shoebox, brick-built buildings which are the airmen's first real glimpse of service life.

'It all depends, not on who your parents were, or what school you went to, but on your performance *now*,' said one officer at Cranwell during my visit. An officer at Swinderby expressed the same sentiment in almost exactly the same words. The RAF is said not merely to tolerate but to encourage élitism; but only the sort based on *professional* accomplishment.

Nevertheless, it is the 'baby budgies' who are seen as the sharpest end of the Royal Air Force. 'They are the *real* RAF, as they would be the first to admit,' I was told drily by one administrative officer at Cranwell; and, as their training costs about £3 million a man, the RAF is understandably keen to give its 'baby budgies' at Cranwell plenty of ostensible leeway for initiative, while keeping a sharp eye on them as they absorb, through Cranwell's fifty-year-old neo-Georgian splendours, the esprit de corps of the Royal Air Force.

Even the airstrip from which training flights take off at Cranwell contains the restraints and incitements of history. It was from Cranwell that the first non-stop flights to South Africa took off in the 1930s, and it was from Cranwell that Sir Frank Whittle developed Britain's first jet engine before the Second World War.

Perhaps one 'baby budgie', at the controls of a Jet Provost training aircraft, had too much formidable history on his mind as he attempted to land on the historic airstrip. Perhaps he came down too heavily. Perhaps he hit a lump of debris on the tarmac. What is certain is that, as the Jet Provost touched down, a tyre burst in a cloud of blue smoke.

The instructor in the cockpit with the student pilot took over the controls almost immediately after the first uncharacteristic bump. He kept the aeroplane (it is *never* merely a 'plane' in the Royal Air Force) more or less in a straight line, and on the runway, as it slowed down.

Within seconds engineers from their nearby hangar had reached the aircraft and replaced the wheel. Within minutes the Jet Provost was inside the hangar. The Corporal engineer quickly produced a thick loose-leaf book in which the aircraft's pedigree was comprehensively charted.

He was philosophical. 'Well, it was due to have its hydraulic pump seen to, anyway. It will take probably a day to put right, three days at most. This was an Emergency State Two – not the most urgent – because there was no immediate danger of a crash.'

'No?' I queried. 'It certainly wouldn't have improved my holiday, if it had happened to me last year at Nice airport.'

'Well,' said the engineer, 'tyres don't often burst. One every two months, I would say. The students may brake hard or hit a bit of gravel. It happens. This guy was having his basic flying training.'

A Squadron Leader nearby in the vast engineering hangar assured me that on potentially highly fraught occasions, there would have been more preparedness: 'During a man's first solo, we bring all the crash crew up to readiness state. The medical staff is geared up and the ambulances are ready. Not to make the man worried, you understand – purely as a precaution. Things seldom go wrong.'

Whisper it softly ('If I hear any more drama about how the adrenalin flows here I shall scream,' said one administrative officer), but life at Cranwell is drama disciplined by the traditions of the service. It is one of the few common factors for all Cranwell men and women during their eighteen weeks of Initial Officer Training; after that, they will go their separate ways, into their highly specialized technical instruction.

It is also a factor conditioned by the splendours of Cranwell itself. The college cannot compete with the imposing and seemingly endless façade of the Army's Sandhurst, nor its vintage (1812). It cannot quite compete historically with the Navy's Dartmouth, which was opened in 1905. Cranwell started life in 1915, as an air station of the Royal Navy, when it was known colloquially as HMS *Daedalus* because the officers and ratings there were on the books of the true HMS *Daedalus*. It became RAF Cranwell in 1919 and a cadet college the following year, when it was still a collection of wooden huts. Things perked up in 1929, when work began on the present central building, which was completed in 1934. Its tall rotunda and dome are now visible on the skyline from miles around – especially at night, when the revolving light at the tip advertises itself as the only 'lighthouse' not run by Trinity House.

The legend is that Naval architects flew all over Britain to find a place large enough (3,000 acres) and flat enough to build an establishment for training officers and ratings to fly balloons and airships; they found Cranwell. It is now a point of pilgrimage for local and regional organizations, including over-60s clubs, whose members take delight in queueing to sit in the padded wooden chair donated by one group of students for the Queen to occupy on royal occasions at Cranwell. Normally, the chair stands in the gallery of the central rotunda, above the vast Great Dining Hall, in which the Royal Standard granted by the Queen is kept in the talons of a ten-foot-tall bronze eagle, while vast portraits in oils of former Air Marshals keep steely watch on today's students.

Cranwell used to be *the* officer training establishment, though there were in fact others in the RAF. Its three-year course, now discontinued in favour of shorter and cheaper expedients, was looked upon as the necessary badge of the élite, a means of predicting in advance who was most likely to reach the top. In the late 1970s, it was decided that all officers would have their basic training at Cranwell, in what is called the 'single gate system', and separate later only as the demands of their specific specialist skills dictated. Thus pilots, engineers and suppliers can *all* feel themselves to be old Cranwellians.

The clear message of the Initial Officer Training course is that, even though a man or woman may require sophisticated technical knowledge, he or she is still expected to have the more primary requirements of toughness and leadership in the field. There are regular sporting contests with Sandhurst and Dartmouth, to underline the point that the RAF must not be inferior to the British Army or the Royal Navy in more macho matters.

Seventy-two periods of the course are devoted to physical education, compared with the 177 devoted to leadership. Leadership, it is clearly indicated, has its physical as well as moral side. Students are told about basic map reading and knot tying, and then put into a two-and-a-half-day field camp. They are instructed in the theory of leadership, in which civilians are involved. Then they are required to lead other cadets in the sort of operations they encountered at Biggin Hill selection centre – moving drums over water with a few poles and bits of rope, and so on. Then there are the more demanding leadership camps at which, for eight days in chilly Norfolk or on breezy Salisbury Plain, the students eat fresh rations, and for six days are taught, by the RAF Regiment, the basics of infantry training that could later help the RAF Regiment in its job of defending RAF bases from enemy attack. The instructors give them arms and blank ammunition to teach them something about arms maintenance and safety.

But none of the armed services lives in a world where there is necessarily an easily identified enemy, who can be overcome with brute force alone. This is especially true of the RAF, whose personnel tend to have more access to secret equipment and facts than the other two services, including access to the highly sensitive nuclear bases which have so often been caught in political cross-currents.

As a Church of England padre I met at Cranwell put it, the RAF can suffer from 'the vulnerability of its community'. A tall, saturnine man with the face of a shrewd and kindly eagle, he was in mufti when I first met him at a sergeants' mess social event, at which officers ritually play sergeants at bowls, darts, snooker and other socially levelling pastimes, including consuming respectable amounts of beer. At such events, where plain clothes are de rigueur, it is far from easy to tell who is an officer and who is an NCO. In a little guessing game I played with a Sergeant who seemed to know everybody, I tried to guess who was who – and was wrong as often as I was right. In other words, I scored just as well as I would have done had I tossed a coin – a fact which tended to indicate that the RAF, perhaps even more than the other armed services, does not have an 'officer type' who can be picked out by his thin neck, horsy patrician face and pink cheeks.

I certainly would not have diagnosed the padre as a clergyman while he wore his brown suit, not even with the help of the glass of orange juice he held in his hand. He told me he had joined the RAF while still young, left it for six years with no intention of going back, 'become organized' in his faith and then gone back into the RAF as a padre. This suited him more than service as a combatant would have done – a fact which I inferred from the understanding way he, the next day, dressed in his RAF trousers and woolly pullover, gently quizzed a class of officer students about how they would reconcile, in their own minds, their humanity with the requirements of the RAF.

It was all part of dealing with the vexed question of the vulnerability of the RAF community, as well as human and moral issues.

'Your country and your allies have been wiped out in a nuclear attack,' he theorized in front of the class of twelve men and three women. 'You are the Captain of a Polaris submarine carrying nuclear rockets. Should you, in those circumstances, fire them?'

Of course, such a question does not have a simple 'right'

or 'wrong' answer. One possible line of thought is that, once your own country and its allies are effectively gone, there is no point in taking any more lives. Another is that it would serve your country, in its slow recovery process, if the other side were taken twenty places back on the snakes-and-ladders board as well. Such questions probe sophistication of thought as well as simple loyalty, and are meant to.

'Well,' the padre asked the class, 'what were the factors uppermost in your mind when that situation was presented to you?'

'That it would be horrible to take all those lives,' said one of the men.

'And do you think that response is one that the Russians in *their* version of Polaris would take?'

A pause. Then: 'No.'

'Yes,' cut in a girl. 'As a human being, a Russian in that position would say there was no point in retaliation, though he might believe in his ideology.'

'Ah!' said the padre. 'What is the difference between a human being and a human being with ideology?'

'I don't think moral values are changed much between East and West,' said the girl.

'Yes, but we started off with the basic functions of a human being, yet we have become conditioned by political ideologies that will influence future action. The Nazis were human beings, yet they were able to exterminate millions of people, because of what they believed. The factors influencing our final decision are a mixture of our status as human beings *and* our beliefs and ideology we have embraced as being ours. Now the Nuremberg war trials: some Nazis said they were obeying orders as military men. What do you think of that?'

An intense young man with metal glasses volunteered: 'In their case, they had no option but to obey orders, otherwise they themselves would be in the gas chamber. In a Polaris submarine, the Captain is not in that position.'

The padre had obviously been over this course more than

once in the past. 'Isn't it an alternative to go into the gas chamber himself, rather than carry out the order to exterminate others?'

It was obvious that the class was rather too young to have considered the possibility that they might ever choose voluntarily to cast away their own lives in such a way. There was a rather mystified silence.

'If you have opinions, you must have the courage of those convictions. A lot of people would stand by what they believe to be right, at the cost of either themselves or humanity suffering. What are the factors operating in the imposition of an ideology?'

'Group pressure and peer pressure,' said the man with glasses. 'The operation of ideology can have a small core; you will get one or two people actually providing the initiative, and other people will then follow on. You haven't got to convert a really large number of people.'

A woman student said: 'The time has to be right to impose an ideology – the state of the economy, or something like that.'

The padre asked about brainwashing. 'You have been brainwashed for the past eight weeks? It is possible for you to be crushed as a person unless you are a strong leader. How can you survive, retain your integrity as a person, and yet fulfil the RAF's expectations?'

A student with a strong Northern Irish accent suggested: 'It doesn't have to be a negative force, brainwashing. It is important to accept the challenge, and see how you can stretch your own beliefs round that, rather than wholly change.'

'Do some of you think you can play a role the RAF wants, and then go back to being yourself again?'

No takers. 'You have got to find a happy medium,' said the Ulster-accented student, after an uncomfortable pause. 'The happy medium between yourself and doing what the RAF wants. You aren't much good if you are unhappy with what you are trying to do.'

It came as almost a shock when, for the first time, the

padre overtly mentioned Christianity. 'There are occasions when I don't know the answer and the only standard is what Christ would want me to do in this situation. I believe individuals to be sacred and unique. By the standards of Christ, our attitudes must be influenced very much by him. There is a tremendous pressure on our military role, and the taking of human life. It has got to be the last resort. Not only being loyal to our side, but being loyal to the humanity of the other side as well. We did not find this common in the Japanese prisoner-of-war camps in the Second World War. The Japanese thought that by the desire to surrender, you surrendered your right to live. They treated prisoners of war as less than human. Now *why* should you care about human beings, why should you be a caring person? Should you, for instance, curry favour with your subordinates by not charging people with offences?'

'No,' said one of the students.

The padre did not agree, nor disagree, specifically. 'You are put into a position to hand out punishments by being an officer, and you need to know the right sort to award. It helps you to impose justice.'

The student with the Ulster accent pointed out: 'There is pressure from outside the service to treat people differently.'

'Yes, like attitudes to homosexuals in society. Were that to raise its head in the service rather than being seen as a disciplinary problem, it would be seen as making the individual unsuitable for service life. There has been a change there. But even nowadays, if you are involved in a divorce, the RAF wants to know if you have been behaving in an honourable way. There has been a mellowing, but not a complete change of attitude.'

A woman student said: 'You have to remember you aren't the most important person who has ever joined the RAF. When you work with other people, it automatically means you care for their problems.'

A ginger-haired male student said hopefully: 'You gain the respect of all the men if they see you smiling and considerate.'

The padre did not comment on this assumption. The student with metal glasses put in: 'If you are able to mix with airmen as well as officers, it brings about a more caring attitude. You can see where people are going wrong, and help them out.'

The padre *did* comment on that. 'Yes, but you aren't carers because it is good for discipline; you are carers because you *do* care.'

If the response to this suggested that the themes were not unrehearsed, at least it proved that some of the 'brainwashing' the padre referred to had concerned the duties of the RAF officer to look after the non-commissioned ranks – duties which, in the RAF, unlike the British Army, cannot be stimulated by the professional intimacy of regimental life, simply because the RAF is not divided into 'families' of regiments.

An older student said seriously: 'Where, before, you were given a book and told to follow it, you are now looking for more intuitive skills in trying to solve other people's problems. You try to give them respect. If you take away problems that bother them, they operate more effectively. A happy worker is a better worker.'

A student with the bluff red face of the archetypal NCO seeking officer status by saying the right thing volunteered: 'You aren't doing it just because of that. Ultimately you are kind just because you are kind.'

The padre displayed no cynicism on hearing his own pronouncement of a few minutes earlier relayed back to him in slightly different words. He merely said: 'Yes, but suppose someone under you is older than your father, how do you feel about that?'

'It is important for an officer to have respect for all people under his command,' said a previously silent student.

From behind the metal glasses came this sentiment: 'If you get respect and teamwork, an officer won't mind a Warrant Officer helping him out, for instance. He won't pull rank.'

'And what happens if you do pull rank?'

'You lose respect.'

'Yes, your role is to be an *enabler*, getting the best out of people with the skills they have.'

The padre turned to the subject he had mentioned to me the previous evening, the so-called 'vulnerability of the RAF community'.

'Perhaps,' he invited, 'you can say what some of the vulnerabilities are.'

A burly young man, who looked rather like one of the well-known British film actors of the war and immediate post-war period, said thoughtfully, 'People aren't quite as stable in their job, if they are in the RAF. People will find it difficult to establish ties with other people. You have a responsibility to make sure people settle into new places and find new roots.'

The padre nodded and reflected that in the Army you could find people who had been with the same regiment for twenty years. At least they were able to relate to a community. In the RAF, people were posted with different families. 'A young couple newly married, shall we say? The girl is from Scotland, and they are posted to Brize Norton. Then the husband is posted to the Falklands. This lovely young girl is sitting in this married quarter area, knowing hardly anybody, and has a husband halfway round the world. It may be best for the Flight Commander to be made aware of the problems, with the suggestion that another wife is sent round, rather than a lusty young corporal.'

The officer students took this point seriously, without even the slight suspicion of a smile. 'In the old days,' said the padre, 'the padre himself did a lot of counselling, but we have taught the skills to the Flight Commanders now ... Now what are the other vulnerabilities?'

Silence. 'Well, if you were a subversive agent, acting on behalf of a potential enemy, and you wanted to undermine the soft underbelly of the community, where would you aim?'

The young man with the intense metal glasses suggested: 'Drink, drugs, anything like that. People with drink problems can be intimidated by subversives.'

'And what happens to the pilot who is flying, and has done something of which he is ashamed and regrets and which is on his mind? Is he able to concentrate on flying?'

After a chorus of ready 'Noes', the padre asked for another example of the vulnerability of the RAF community, of which officers should be always aware. He supplied his own answer: teenagers. In Germany there were youth workers to help handle the problems of rebellious adolescents, and these could help sort out many different types of potentially damaging situations.

'I had a case at my last station where a mother came to see me,' remembered the padre. 'She said she hadn't seen her teenage daughter for three days – and three nights. I had red alerts flying in my mind. I got the feeling there was more to this than met the eye. I rang the youth officer, and he said I should speak to a certain social worker. Within an hour, I was able to take the mother to where the daughter had been staying. Who do you think are the caring agencies to be approached?'

It was predictable that some of his audience would shout out, 'The padre!' And they did.

Yes, agreed the object of this confidence, *if* you could find a padre: there were only one hundred chaplains in the entire force, covering all denominations. He agreed that medical officers would be a good idea, because some people had to be sent away for psychological reasons: in Berlin nineteen RAF personnel had been sent home in two months because of psychological factors.

'Policemen,' suggested a student.

'Yes, people don't think of the policeman. But I have known of many policemen who have put drunks in cells for the night and, in the morning, kicked them in the backside and reported nothing. Don't forget policemen are human beings. Preventative care is better than disciplinary reaction. The police have to maintain the law, but if it is possible to interpret it in a caring way, they often will.'

Whether policemen who kicked drunks in the backside were everyone's notion of a caring person is something I thought it best not to raise. In any case, the padre was turning the dialogue towards the way *not* to be caring, and a way that was possibly alluring to a young officer just feeling his feet.

'Suppose you have a Warrant Officer with personal problems coming to see you, what then?'

'Imagine it is your dad.'

'Right! But do you sit down and solve his problems for him?'

An officer student with a broomlike moustache and confident manner dismissed the idea. 'You want him to solve his *own* problem. If you just ask him questions, he can talk it out. If he does something you recommend, and it doesn't work, you lose your credibility.'

'Suppose you give him advice and it *does* work?'

'He becomes dependent on you. The object is to make him independent.'

'Yes. I remember a situation in Germany, where the banks are very happy to lend money to you. An airman came to see me, and told me he was in a financial mess. That is a situation that could be exploited by a subversive agent, especially in a situation like Berlin . . .'

I left the classroom aware that being an officer in the RAF cannot be simply a matter of leading men in hand-to-hand combat in the field, and displaying the right manners in the mess. There must be a high degree of awareness of human identity, and of human vulnerability. To this end, the RAF appears to utilize the institution of the padre in officer training very skilfully, using him as a modern rather than a conservative churchman, concerned about social problems, and at home with their special vocabulary.

The Director of Initial Officer Training at Cranwell, a Group Captain, told me cheerfully that all this training could do, with or without shrewd padres, was to 'put down the first layer' of being an officer in the RAF.

It is a first layer which emphasizes that, though he may regard himself as a trained technician, the RAF regards a man (or woman) officer as part of a tough fighting machine.

'Every officer,' said the Group Captain, 'must have the appropriate degree of leadership, must be able to communicate effectively and maintain certain standards of dress, deportment and behaviour, *as well as* having a knowledge of defence-related subjects like the nuclear deterrent and the relative strengths and weaknesses of NATO and the Warsaw Pact.'

But surely the concept of the 'officer and gentleman' was pretty dead in a highly technological Royal Air Force?

'No, it isn't,' I was told. 'The Royal Air Force certainly doesn't want Colonel Blimps, but the officer-and-gentleman thing is still important. I see us as being pillars of a mature society. I happen to be a believer that the pillars, in any society, have a certain responsibility which is *social* as well as practical. It is an unwritten thing, but we do not expect Margaret Thatcher to roll up at the House of Commons in sneakers and jeans. You would expect her to be able to conduct herself in any sort of society, and to be proud of her when she sets out to make a speech. It would be very difficult to say what direct value these factors have got; you would find it difficult to come up with a pat answer, but we do try to engender that sort of understanding in our officers.'

The Director of Initial Officer Training said he was about to challenge the existing syllabus, so that it would allow more attention to the sort of social polish he valued. 'We don't do it directly. We embarrass people into it.'

Embarrass? How? 'There are certain dress standards. If a chap isn't dressed in the appropriate way, I believe that, by a process of metamorphosis, he will look around him and feel uncomfortable; and, if he shrivels, it is the best method. We need to create the sort of environment and structure which will give you an opportunity to allow these things to happen. For instance, we are trying to introduce a dinner night, where they will have the customary meal – which will cost them more – and they will wear dinner jackets, and we will get a high-quality speaker from industry or politics to talk to them. Afterwards, into the bar and chat with the speaker. That type of occasion, together with guest nights, *and* the sort of building Cranwell is, all hit the right note. I abhor the idea of teaching people etiquette; but, through the atmosphere here, we can try to do things in a subconscious manner. General standards of behaviour can be brought about by a process of osmosis which will soak into them. Following others' example. Most of them come from good families, after all.'

At first sight, this argument struck me as one more likely to be heard, let us say, in the Grenadier Guards than in the highly computerized and socially egalitarian Royal Air Force.

The Director made haste to explain what he meant by that phrase 'good families'. 'I mean a family with good moral values, that's all. We have chopped the odd guy for lack of integrity.'

There had, apparently, been a recent case. The man was on the fat side, and didn't like physical training. He had told a member of the Cranwell physical education staff that he couldn't do physical training because he had a dental appointment. The Flight Commander had rung the medical section, and been told there was no sign of any such appointment and that the man hadn't been there. Tackled later, the man swore he had in fact gone for the alleged appointment. Only under pressure did he agree he *hadn't* had a dental appointment.

'I am under a lot of strain,' he explained, 'because my grandmother has just died, and I am trying to keep it all to myself, because I don't want to be re-coursed.'

This sounded plausible, as no student likes to be directed to take the course all over again. But when the Flight Commander telephoned the man's home, he was told that the grandmother had indeed fallen down and bruised her leg, but was otherwise in good health and spirits.

'We considered that chap was beyond the pale,' said the Director of Initial Officer Training, firmly. 'To be fair, there were problems. His father could not afford to let him go through university. He had had a local authority grant;

and, when he had left university, had been under pressure from the local authority to pay it back. Still . . . that sort of chap can't apply again for officer training. Fundamental integrity is not something we can instil in eighteen weeks.'

The Director made it quite plain he was in favour of the 'more socialistic' 'single-gate' entry system to the Royal Air Force, which had taken over from the old system, in which the given percentage of the officer intake who happened to go through the Cranwell three-year course were regarded as the backbone of the RAF. That was élitist and divisive, said the Director. And it would also be divisive if women were treated differently from men in becoming RAF officers.

In fact, women at Cranwell do almost exactly the same fieldcraft exercises as the men, with the exception that, if there is something heavy to be carried, a thirteen-stone man may be asked to carry it rather than a six-stone girl. The women sleep in the same ten-by-ten-foot tents on these exercises.

'They are so well parcelled up in thick clothes and sleeping bags that there is not much inducement to unworthy conduct,' said one senior officer. 'They are totally integrated in barracks with the blokes. They use the same showers and baths. The girls like it, and wouldn't do it in any other way.' (Whether the girls passing through Swinderby School of Recruit Training – the 'Cranwell' for non-commissioned ranks – would like the same arrangement or not, I was later to discover there that, in practice, the regime was less reliant on individual common sense.)

I asked the Director if the presence of women, in such proximity, had a discernible effect on the young male officers. Yes, it obviously did have an effect, said the Director. The men liked having pretty girls around, so it was good for morale. But it took the men no more than the first two days, out of the eighteen weeks, to realize that women were just part of the scenery. But possibly the fact that the women tended to be very mature in their outlook, with great character and determination, exerted some influence which lasted longer than that two days.

At the time of my visit to Cranwell, there were moves to enhance the maturity of officers by adjusting the time their initial officer training ended. The system, at the time, was that the students were well and truly put through it for sixteen of the eighteen weeks. Only in the last two were they expected to behave like people who very shortly would be *giving* the orders.

'We are moving towards a system in which he has only twelve weeks of beasting, after which he will be required to *apply* it,' said the Director.

Such a change could well make it easier for some students to graduate from their Initial Officer Training course in the splendours of Cranwell *without* having to be re-coursed and thus having to do the last twelve weeks of the course twice.

One student, who had at last succeeded after being re-coursed, told me that it was indeed the inability to *exert* rather accept leadership that had, earlier, been his undoing.

He was a twenty-six-year-old engineer who left grammar school at seventeen for a technical apprenticeship in civilian life, which took him four years and left him with a certificate in mechanical engineering. He then settled for white-collar work as a car salesman, which he found no more satisfying than the humbler tasks of basic engineering.

'I wanted to come into the RAF because I was looking for a solid career I could really get involved in,' he told me. 'Something worthwhile. Selling cars was a dead-end future. I didn't know what I expected at Cranwell. There was more emphasis on academics than I first expected. But that proved not to be the problem: it was well planned and structured. But the course here is very leadership-orientated. I think I had a lot of the *components* of leadership, but when I tried to fit them together in a stressful situation, I did not have the full impact I was required to have. The confidence didn't come over, and the leadership tended to dwindle.'

I asked him *why* he hadn't had the confidence. Because, he said, he hadn't settled into the atmosphere of Cranwell quickly. It was 'a bit foreign'. It had taken him a little too long to understand what was required of him, and to get on with it. His re-course was still a disaster in some ways, especially from the point of view of the vital confidence. There were twenty-six of them being re-coursed, all conscious that they were lacking in confidence. After the first four weeks of the re-course, confidence had started to flow. They had a fortnight of adventure training in Scotland. 'It worked.'

In this engineer one could see very easily the sort of technologically minded person the RAF wants, the RAF needs – but also the sort of person who might well have the most difficulty with the ways of a disciplined armed service. The compensation for the difficulties such people face is that very often they find that, having instilled military qualities into their *job*, their *private* personality becomes more confidently easy-going, not less. They do not feel the same pressure to exert themselves in the dining room or the pub.

The engineer reported that his wife in particular had commented that he was now a much nicer person. 'I tend to look upon what *other* people need a bit more. I am less selfish. It is not a face which I put on for the course, it is an underlying thing. You stand back and look at things a bit more, rather than diving in headlong and giving offence. The social skills package at Cranwell is good. You change almost imperceptibly, as far as you yourself are concerned. In company, say, I am more careful I don't come out with nasty little catchphrases, censorious catchphrases, like I tended to. Many people said I had an annoying personality. I could get on people's nerves. Quite possibly I was self-righteous: a little bit too keen to impose my own ideas on all situations. If a couple were discussing something, and I had strong views on it, even in just a social discussion, I would definitely let them have my views. Now I think, "Is it necessary, is it worth it, what am I going to gain, apart from putting their backs up?" This is a great advantage, and it was because of Cranwell.'

His was a fairly typical engineer's response to Cranwell –

from what one could describe as the ground-officers' class or caste within the RAF. I discovered more evidence – from engineers doing their advanced engineering management training at Trenchard Hall, an imposing building of civilian aspect in the Cranwell acres which is named after Marshal of the Royal Force Viscount Trenchard.

Engineers, one twenty-five-year-old student told me, were used to working on their own responsibility and even on their own on RAF stations once they were out of training; aircrew tended to work and play together. This was reflected in training itself.

'Baby budgies?' he said. 'They are told they are what the Royal Air Force is all about, and that they are important. Baby budgies have an opinionated view of themselves, and inflated egos. Once they have had a few failures they are better for it.'

Couldn't the suppliers feel *they* were being regarded as second-class citizens by the engineers when the engineers referred to them as 'stackers'?

'No, because they bite back as well by calling us blunties, and it is all in good part. They *could* accuse us of snobbery, but they don't – because we don't act in the same manner towards them as the pilots act to the rest of us.'

I asked the student if such irritations ever made him regret coming into the RAF. 'No, I don't think any of us would say we wished we hadn't come to Cranwell. Just childish clashes with senior officers, like when we wanted a Land Rover for an exercise, and the boss wouldn't let us have it because, he said, he needed it to visit another boss. He didn't have an overall view of the picture. It was a form of pulling rank, I suppose. We are supposed to be officer students, but are treated as student officers – when you get as far as here you have already been commissioned, but officers are still treated like cadets. Sometimes they seem to treat you like children on the course – and then they expect you to go out and do a management job over eighty men. They seem to have a double standard in that respect, but I suppose that would be so in any organization.'

It is easy to see that in some, almost inevitable, respects an institution like Cranwell must perform a balancing act between military requirements and the sophistication of the techniques being taught, and the officer students learning them. It must apply especially with engineering management, where a civilian psychologist with sixteen years' experience in the RAF, and eight years working in the social services and social services management field, talked about interviewing techniques, understanding skills, dynamic understanding and the Egan Model.

The head of the Engineering Management Wing, a Wing Commander with scholarly glasses, rescued me from this formidable jargon by explaining that one of the difficulties faced by young officers was getting information from the men and women under them. Subordinates tended to be of two types: those who rushed in and wouldn't stop talking, and those who wouldn't volunteer anything, but had to have it drawn out of them like pulling teeth. Officers in engineering management training faced situational exercises, in which they were rung up at short intervals with simulated problems, which they had to solve by telephoning other people. The problems can vary from a missing wife to a complaint that a car with a VIP inside has been officially stopped by a military policeman at the rear of an aircraft, with the result that the VIP had banged his head on the windscreen.

Yes, one can understand the feeling of the engineers, suppliers and administrators that the 'baby budgies' have comparatively simple lives in their cockpits, being told they are the salt of the earth, and never having to *manage* those things which are more unpredictable than aeroplanes – human beings. But, as the Assistant Commandant of Cranwell, an Air Commodore with previous high-powered experience at the sharp end, told me: 'I stick to the view that we are an Air War Force rather than an Air *Staff* Force.' In other words, the RAF is not about desk work, nor even the sophisticated management of its engineering, but about piloting aircraft that win wars.

When I visited the Flying Training School at Cranwell, the air of enhanced excitement was apparent.

The very start of the pilot's practical training is in the Jet Provost instrument training room. This, I found, was a room about the size of two drawing rooms, almost entirely filled by mock-ups of Jet Provost cockpits. Student pilots were at the controls of cockpits, which tilted and dipped as they would do in the air. The students stared ahead at opaque windscreens (no input from anything except the instruments) and did their best to fly over a given course, while hazards were progressively introduced by an officer who was ostensibly 'back at base', but was, in fact, sitting a few feet away, on a wooden chair beside the cabin. He was watching a map, on which a mechanical fibrepoint recorded the course being followed by the pilot in the cabin.

While I was watching, a student barely managed to turn away in time to miss a civilian air route. I began to feel, even in such a crude simulation, the importance of the pilot's craft: only he has in his hands not only £3 million worth of equipment, but also many hundreds of lives. The exercise finished, the student told me that the feeling was quite different from real flying, that even the controls felt different – more stiff and heavy – but that the training in *procedures* was none the less valuable. It was important, once you were under real stress, for the procedures to be almost automatic.

When I spoke to a group of young student pilots who had just been involved in real flying, it was a rather one-way conversation. As I struggled to keep pace with my notes, I understood what one of the pilots meant when he said: 'Once you start to pilot an aeroplane around the sky, if you are not up to the speed of the aeroplane, you get behind the jet – and things start to happen you can't cope with. Things are being injected into you faster than you can cope with them.'

Touché. The other comments came thick and fast.

'You must always follow the basic rules of trimming the aircraft.'

'Whatever flying you have done in civilian life, this is different. You can crash-land a civil aeroplane in a field, but if you try that with a jet –'

'You will be happily flying along with the instructor, when he tells you of an emergency; and, at the same time, Air Traffic Control is talking to you. And when you think you have got on top of those situations, the instructor will give you *another* emergency. As the course progresses, it starts to get a bit tense and –'

'You need airmanship, awareness and co-ordination. Some people say all this is a lot easier to handle than a car, because there isn't so much traffic about, but actually it isn't and –'

'It doesn't even make you a better car driver. On the contrary. Some students tend to be a little cocky and aggressive which –'

'You can walk into a squadron car park at RAF stations in Germany, and see the Porsches and the BMWs lined up. All the pilots have the Porsches and the rest have the BMWs. A pilot likes to live it up in every facet of his life, so –'

At last I was able to get a word in. Did any one ever feel he would *never* get on top of the fast-moving life of the pilot?

'You don't think about two years' time, that's the truth of it. You think about tomorrow. Tomorrow you will be doing a solo flight. You don't have big goals, you have step-by-step ones and these –'

'It is quite overawing if you think too much about it in advance. With a Harrier you want to fly, there is a mountain ahead of you, and you can only take it in short stages. If you *think* of yourself as a Harrier pilot, and you haven't even gone solo yet, it is completely unrealistic, completely –'

'I still want to fly a Harrier myself. I like the aircraft itself. I like the idea of being able to land where you want to land. It has got such a reputation. Well, I wouldn't know if it's a realistic ambition, because I haven't gone solo yet, but a Harrier is different from –'

'The old days of the war! Aircraft cost a lot more today. It is worth spending £3 million on training me, if I am going to be given £16 million worth of kit at the end of it.'

The only student pilots I encountered who were *less* effervescent, and who spoke at less than four hundred words a minute, were those who had been forced to the sidelines, temporarily or permanently, by sickness. It is a high life for pilots, but a vulnerable one.

One student had been off sick for a nasal polyp which had been removed, leaving him fit to fly again after his convalescence. Another student, aged twenty-two and determined to join the RAF from the time he left school at nineteen, was not so lucky. He discovered a sinus problem, which meant his ears didn't clear properly on or after flights. He had been away from flying for three months; and very doleful he looked. He had already done thirty-seven hours' flying before his problem was diagnosed; the following week the ear, nose and throat specialist was due to tell him whether he could fly for the RAF or forget the whole idea.

What if the RAF said he couldn't fly? 'Oh, a shattered dream and all that,' he said with a reassertion of the archetypal pilot's jauntiness. 'But if they say I can't fly, I will accept that, and apply for a different branch of the RAF.'

Yet another student, aged only twenty, had already done thirty hours in a Jet Provost when he was discovered to have glandular fever: not only his future flying career, but perhaps his whole career in the RAF, was at risk. Perhaps pilots *need* to be jaunty: if they were too reflective, they might see too clearly that their whole lives are as physically vulnerable as are those of the concert pianist, the prize-fighter or the champion ice skater.

Perhaps the pilots *need* their self-supportive feeling of being special. Perhaps, the head of the flying school told me, British air defence needed it as well.

'In NATO exercises,' he told me 'the RAF may or may not get the best ratings for the aeroplanes, but we achieve excellent ratings for our pilots. The Americans seldom get "excellent" ratings, which the Brits do. The Russian fighters do not yet match the American ones in advanced technology, but they are catching up fast. The Soviet Air Force do not yet have as much operational practice as the RAF, but the situation is changing, so we must go on trying to improve our operational tactics and methods.'

It was, in a nutshell, the justification both of Cranwell and of the 'baby budgies' who sometimes so irritate the more measured, less effervescent RAF classes, including the Other Ranks.

13

THE REAL CLASSES (2): THE AIRMEN

'An officer wanted some cigarettes and said to the new recruit, "Get me some Benson & Hedges cigarettes. If they haven't got Benson & Hedges, bring me anything." An hour later the recruit came back – with a Mars bar. That was the most stupid example I came across.'

– Flying Officer on the staff of RAF School of Recruit Training, Swinderby, Lincolnshire.

'I had to do a discharge interview this morning. The recruit said he just didn't like service life. After seven *days*? I explained it would not be like this once he got outside into the RAF proper, but he had made up his mind. A pity. Perhaps he was looking for a father-figure in the RAF and didn't find one – but I don't think he'll find one outside, either.'

– Women's Royal Air Force Squadron Leader on instruction duties, Swinderby.

'Don't put "RAF" in big letters on your luggage,' the twenty-three-year-old Flying Officer warned a batch of new airmen recruits. 'You could pose a threat to yourself and your colleagues as well. You could be recognized by these terrorists or subversive agents. And if you go down to town on Saturday evening and you are sitting in a pub, what mustn't you do? Hodgkinson?'

'Get drunk and start talking, sir.'

'Right. One of the best ways to advertise yourself is to start blowing your mouth off. It's the best way of advertising yourself apart from short haircuts and the fact you look neat. Beware of that. And don't go to the same pub every night at the same times. Because, Wheeler, because . . .?'

The days of the simple airman who was there simply to do as he was told (if they ever existed in the era of high-technology flying) are definitely over. Neither the technology nor the lines of battle are now simple: computers and word processors have complicated the one; the terrorist and the saboteur, operating in 'peacetime', have complicated the other.

At RAF Swinderby, in flat bleak Lincolnshire, home of the RAF School of Recruit Training, 6,500 or so young men and women go through their initial six weeks' (men) or five and a half weeks' (women) basic military training. Long-serving officers and NCOs there say that the education in matters military they have to give has changed greatly in the past ten years: alert RAF men and women must now always be aware that the enemy will not necessarily be an apparent one, that the very real carnage of war can exist in times of nominal peace, and that in this battle everyone has his part to play.

Wheeler (names, as always, fictitious), sitting in the lecture room at Swinderby, was possibly not the brightest of the sixty students of the training flight I saw. They were all aged between seventeen and twenty-five, and had gathered in the lecture room on the fifth day of their six-week course – a course designed to turn even youths who have lived in jeans and trainers, even those who have perhaps successfully avoided the dentist for years and the hairdresser for months, into presentable and aware basic airmen.

Wheeler struggled valiantly to remember why he shouldn't go to the same pub at the same times. He was one of the youngest of the training flight. The barber had tamed his hair to manageable proportions, but the pimple on his cheek and the blackhead in the centre of his forehead had escaped disciplinary action. He scratched what was left of his hair.

'All right, Smith, you try.'

'Because terrorists will know where you will be at such-and-such a time,' said one of the more mature recruits, his moustache well trimmed. 'You will be an ideal target.'

'Correct. You hear that, Wheeler? Remember that. Why must you always label your travelling bags with your *name?*'

'So you don't advertise yourself, sir.'

'Are you listening to the questions, Wheeler? Stupid answer, wasn't it? If someone sees your bag there, and it has your name on it, they can call you to it before someone thinks it is a bomb and blows it up, can't they? Now why shouldn't you leave your bag unlocked for ten minutes while you make a telephone call, Wheeler?'

'Because the terrorist can dump a bomb inside the bag, sir.'

'Excellent, Wheeler! You get on the train in London after calling home and on the way to Lincoln the bomb goes off, killing you and a lot of other passengers. Now what about your car? Johnson?'

'Make sure there is not something suspicious about the vehicle, sir. Make sure someone hasn't tried to force the lock.'

'Yes. And make sure there aren't any suspicious wires hanging down. If your posting is to Northern Ireland, you would have to do that every day. Be sure you lock your car. Keep it in a garage if you can. The thing to take away from these courses is that security is *everyone's* problem, not just senior officers'. There is a constant threat to civil security and also *national* security. This is your responsibility as well, because one day when you come into touch with any information of use to a foreign power, you become a security hazard. You with me?'

A chorus of 'Yes sirs'.

'We'll see,' said the Corporal working alongside the Flying Officer at the rostrum. 'Say I am on leave this weekend and am sitting in one of the locals when someone comes up to me and buys me a drink and tries to get information out of me. What would I do? Miller?'

'Say absolutely nothing.'

'Absolutely right. And then report to the police so they can investigate it. It could be one of these KGB people. You will say nothing, you will avoid questions and you will report to us when you get back. Quite simple.'

'That's about it for today,' said the Flying Officer whose boyish face was at odds with his incisive voice and authoritative manner. As the recruits filed out of the lecture room with studied discipline, he turned to me and explained: 'Remember they are only on their fifth day. Training in this respect of security has changed greatly in the last few years. It is more thought-provoking. Giving them a degree of awareness is certainly my prime concern. We must give them some idea of what is going on around them. There are twenty of them with no previous contact with any of the services. These people have got to be taught everything, including what the different ranks are and what their kit is for.'

The youth and simplicity of many of the new recruits were obvious as I watched a new batch of eleven arrive in a coach from Grantham and proceed at the double into the hangar where they were given their first taste of military instruction and drill. I asked to speak to the two who appeared, at a superficial glance, to be the most likely to give a coherent view about what they hoped for in the RAF; it turned out that *both* had in fact some family knowledge of the RAF – a factor which, say the instructors, is pure gold.

One of the two was sixteen, and came from North Yorkshire. He wanted to join the RAF because it was a good career with prospects. He wanted to become a mechanic and eventually a technician. His uncle was in the Army, and his grandfather had been in the RAF. Yes, he would like to become an officer in due course.

The other recruit I picked because it is comparatively rare to see a black face in the RAF. He was nineteen but looked older, a firmly spoken young man who, for a newcomer with about ten minutes at Swinderby, seemed already at home. His father, who had come to Britain from Jamaica, was a driver. I asked why he wanted to join the RAF. 'My father was in the RAF and the RAF is home life to me. I want to be an air technician, electronics. I was a boarder at a school in the Midlands.'

I asked him how friends of his own age regarded his wish to join the armed services – there had been a time not so long ago when 'peer group pressures' had been all against it, especially among black and Asian people. 'I have got three brothers in the RAF, I know three other guys in the RAF as well, and all the others think it is a good idea, too. No problem.'

Other recruits, without such background support in the ways of service life, *can* have problems.

Some particularly dim cases have become part of the folklore of Swinderby. One recruit was asked by his Sergeant to get him a Valentine card from nearby Newark. 'Take my bike,' said the Sergeant. Over an hour later, the Sergeant began to wonder what had happened to the Valentine card and, for that matter, to the recruit. It was another quarter of an hour before the Sergeant, looking anxiously out of the window, saw the sweating recruit appear up the road, *wheeling* the bicycle.

'What took you so long?' demanded the Sergeant.

'You said take your bicycle and I can't ride one.'

Another recruit attempted to endear himself by claiming he had a degree in Mandarin Chinese. A Corporal challenged him to speak some. The recruit launched into a tirade that lasted what seemed like several minutes.

'Now tell me what it means,' said the still dubious Corporal.

'It means, "Two number forty-sevens and a bowl of sixty-five, please."'

What about the other extreme? Are there ever recruits presenting themselves at Swinderby who should really be offering themselves twenty miles up the road at Cranwell officer training college? Very occasionally, apparently. One Corporal told me: 'I have never met any recruit of whom I would say, "He ought to be at Cranwell *now*." But I have seen a few who, in a few years' time – three or four years – might be suitable. If a person is good enough for Cranwell, they would be looking in that direction for themselves – although there are people who have the ability to be officers but do not want the responsibility.'

Most recruits are neither Cranwell material nor absolutely hopeless. One Corporal I met in the NCOs' recreation room of one squadron at Swinderby admitted candidly that recruits were usually keen to start with, but after a few days of drill they became 'fairly bemused'. He blamed it on 'outside life' which was not gearing young men for the demands of the fighting services.

'They have a lot of problems with boots,' he said. 'Civilian fashion is often for soft shoes, but it doesn't exactly strengthen the feet. The best way to wear in boots is to get on with it and wear them; but they are not very keen on that. You have got to pull them through everything.'

The Corporal remembered quite well his most difficult case – a young man so nervous that he 'went into manic mode' every time he was spoken to. He always looked as if he had pressed his clothes with a brick: 'They find it isn't what they expect. It isn't just a job, a nine-to-five job. It is a way of life. It is clearly a shock to them when they realize just what it is.'

Another Corporal with a longer experience of Swinderby was more merciful in his estimate of recruits. 'The majority of them are quite young in their outlook on life. The thing is not to give them a lot of time to think about homesickness. We don't turn our backs on it and pretend it isn't there. If they do suffer from it, we tell them first to talk about it among themselves and only then come to us. It hits them when they go home for the first time after one and a half weeks. Then they come back here thinking that it was nice being at home, not being shouted at.'

The Corporal said that the most satisfactory recruits were usually ex-servicemen or ex-cadets – though the one with the most difficulties he had ever come across was an ex-

Naval man who, because of his service experience, was made senior man in his flight at Swinderby. After about three weeks of training, he started to show signs of stress, to the point where he was not sleeping. He used to prowl around the barrack block in the early hours of the morning. If anyone he met was going to the ablutions in their night attire, he would ask for their identity card and, if told it was not kept in their pyjamas, would tick them off or put them on a charge.

It all came to a head one morning when he started to shake and shiver when being interviewed on his state of mind by the Flight Sergeant and a Corporal. He was taken to the psychiatric unit where he confessed that his wife had been finding it hard to cope in his absence; so that, on top of his responsibilities as senior man on his flight, he had family tensions as well. After being told not to be quite so intense and serious, he slowly began to cope, and passed out at the end of the course.

The more usual picture of a recruit who finds it too difficult to cope was presented in his present intake, said the Corporal. This was a young lad of seventeen, who looked and behaved even younger than his age. He was a typical mother's boy – his mother had done everything for him, including dressing him. He had bought himself out – the majority paid about £95 to do this, though those who joined at under seventeen would get a free discharge at eighteen.

I asked a black Corporal why he thought there were so few West Indian and Asian recruits presenting themselves at Swinderby. He said he was not West Indian but Nigerian; his father had been in the Nigerian Air Force. This, and the fact that he had previously been with the Royal Marines, had given him a military background denied to most West Indians and Asians. He had one West Indian in his flight at the moment, and he was not really a very good recruit.

'Since I have been here – three years, in the course of which I have handled eighteen flights – out of the thousand recruits I have seen, I suppose only ten have been Asian or black. They are not usually orientated to the services. They are not so much frightened of getting aggro from their white colleagues: it is just that the services life is not part of their vocabulary. *My* friends don't understand why I want to have anything to do with the services. It is mostly West Indians we get – and they are very probably very anti-Establishment now. They are laid-back and their whole culture doesn't appeal to discipline.'

So said the Nigerian Corporal, who admitted that when he had first joined the Royal Marines he had a chip on his shoulder. Within six months that had disappeared. He had been pressured into situations which forced him to become 'one of the boys'. You had to prove yourself tough, or get ditched, he said. 'To become one of the boys, you do as the Romans do when in Rome, basically. Five years in the Royal Marines gave me a lot of knowledge and toughness. The RAF is a totally different service from the Marines. It is in a way easier-going. In the RAF I would still do what I was told, but I would think about it. There is a much more individual discipline in the RAF. In the Marines, discipline was imposed on you. You have got to be individually motivated in the RAF.'

Officers say that spending six weeks at Swinderby is not long enough to bring about this motivation completely but that it helps.

The ritual is fixed. The first two days of the course are the reception phase, when the recruit is kitted out and has any medical problems attended to. During the first weekend he is introduced to gruelling drill and RAF general knowledge. He learns how to keep his barrack block clean, how to pack his bed, polish the floors and attend to his personal hygiene. After the weekend he has to run one and a half miles, which he has to complete in under twelve minutes or face remedial physical training until he *can* do it in twelve minutes.

On the second week, he gets his weapons training – self-loading rifles (SLRs) for the men, sub-machine guns (SMGs) for the women. On the third week there is more advanced weapon training and lectures on internal security, followed during the fourth week by a three-day camp under canvas. The fifth week sees the recruit preparing for the graduation parade in which – he hopes – he will take part, and doing three periods of arms drill a day, so that he doesn't drop his rifle at the march-past. On the sixth and final week, the individual flights in training come together as one intake, so that they can co-ordinate properly at the passing-out parade.

This little scene reads more easily than it plays. With an average of two 'O'-level passes and some good CSEs, the recruits arrive at a varying rate during the course of the year – 5,500 RAF recruits and 600 members of the Women's Royal Air Force. The middle of July, school-leaving time, is the peak. Each RAF flight has sixty recruits; the WRAF flights have fifty members. Men and women get lectures from padres on personal relationships as well as religion, and women get more lessons in 'general service knowledge' than men, because it takes into account what are euphemistically called 'women's problems'. Women learn how to *fire* weapons but do not have lessons in arms drill: all have a 6.15 a.m. reveille, first instruction at 8.30 a.m. and lights out in the barrack blocks at 11.15 p.m. There are rigid checks every evening to see that everyone is in bed.

Airmen and women recruits are not trusted quite as much as are young officer students at Cranwell. There are separate barrack blocks for the men and women, and there are separate training flights. Officers explain that if the girls were in the same flight as the men, they would 'feel left out', a line of reasoning that even on repeated inspection has been unable fully to reveal its pearls of wisdom to me. The girls appear to be almost as tough in their basic approach as the men, though it may be true that, age for age, they appear more *mature* than the men, and might find the young men's high spirits a little fatiguing.

Certainly, though the women fire only forty-five live rounds from their sub-machine guns at the range compared with the men's fifty-five, they are the equal of the men in their willingness to pick up and use a loaded gun – one, too, that can be even more devastating than the men's self-loading rifles.

One woman Squadron Leader at Swinderby, a lady whose soft voice and somewhat ornate gold-and-pink spectacles did not conceal the incisiveness of her manner, told me she had encountered only one girl who had refused to pick up her sub-machine gun since weapon training had been introduced in 1984: 'The girls generally enjoy their weapons training and are generally very good at it. We insist on girls being arms-trained. But this girl just refused to pick up her gun. She suddenly looked down at the gun and realized she didn't want to be part of an armed service, though she already had live rounds up the spout to fire at the cardboard figure of an enemy. She had moral qualms and decided to leave. Normally the careers information offices do a good job in warning girls about weapons.'

The RAF insists that all except those destined for the medical trades be arms-trained. In practice, even the medical trades voluntarily agree to be arms-trained. What if they did not agree? 'If the medical trades said, "We don't wish to", we would not make them. But the situation hasn't happened – they have all said, "Yes, we will go ahead",' said the woman Squadron Leader.

The most rigid caste lines at Swinderby are not between men and women, not even between officers and airmen, certainly not between those who have the 'right' and those who have the 'wrong' accent. It is between those who can cope and those who sadly cannot; the real division does not express itself in sex, social or even educational terms.

'Beasting' is an unpleasant word, but it is a word still used to describe what is done to the recruit in the first weeks of his training at Swinderby. A Corporal will shout at recruits and turn on the heavy menace to such an extent that the Sergeant he works with has to balance this with a little fatherly understanding.

I spoke to one Corporal who was practically having to reconstruct his own personality because, instead of working with a Sergeant on his flight of sixty recruits, he was being forced, through illness, to take the flight on his own.

This was his version. 'My role has had to change. We usually do the hard man, soft man approach. It is usually the Corporal who will shout and coerce and the Sergeant who then stills the waters a little bit. Now my approach to my flight has to be completely different. It is no good *both* of us going in there beasting them – someone has to be approachable. So I have to be approachable now I am on my own. I have to manipulate my role, so that I am softer and yet firm. This is difficult to adjust to. It is 90 per cent planned and 10 per cent spontaneous. I don't lose my temper at any time, but we will have hard days and soft days.'

Soft days are when the recruits in the flight need building up after being beasted to near the point where they are in danger of thinking, 'I can't do anything right, so there's no point in me trying.' In the first ten days the Corporal and Sergeant have plenty of soft days; after that the hard days increase – but will not go on to nearly the end of the course, or the morale of the flight will be inadequate to get them through their final grading. Recruits are offered a 'carrot' by being told they can have a night free if they have achieved a certain objective. The argument is that if you give the recruit something, and then take it away for bad performance, it is more likely to get results than if they are not offered concessions at all.

'It's a little bit of psychology – no, call it experience rather than psychology,' said the lone Corporal.

Despite the difficulties of running a one-man band, the Corporal found that only one of his recruits had to be reflighted. He had been a difficult case. He was charged with having a dirty kit, but that was only the start of it. He was marched in front of the flight commander and awarded three days' restrictions, colloquially known as jankers. Whilst on restrictions he got picked up wearing the wrong shoes – he was wearing his number twos instead of his (best) number ones. He was charged again with failing to comply with an order. He got a further five days' restrictions. Still the sorry saga was not over. After a full kit inspection in which his own was faultily displayed, he was charged again.

'His problem is his attitude,' said the Corporal. 'He is *capable* of doing it correctly. Having had a bollocking for doing something wrong, he will maintain the standard for a couple of days and then lapse back again. Generally, their worst stumbling block is settling down in the first ten days. We get more people applying for their discharge in the first ten days than at other times: if we can hold them for the first ten days there is a good chance they will stay.'

The two-and-a-half-day field course at North Luffenham near Grantham has acquired myths that put some new recruits off. 'Flights come back from there and tell war stories to the people who have yet to go,' said the Corporal. 'You then get recruits going down there with a negative attitude. We tell them not to listen to all the stories they are told; but they still do listen to an extent. I have never taken a flight down there without someone wanting to pull out.'

This despite the fact that conditions at training camp might be regarded as opulent by the Royal Marines, who sleep in bivouacs on their exercises – simple plastic sheets tied to trees a couple of feet above the ground. I shared these with the Royal Marines when researching my book *Ruling the Waves*: they are testing but not unendurable. Airmen recruits in the RAF do not have to endure them, except in a limited way in the summer months. Otherwise they occupy ten-by-ten-foot tents, a rather more civilized form of protection.

Did that mean airmen were softer than Marines? The Corporal was frank. 'Personally, I *know* they are softer than the Marines. We are training people to do an entirely different job from a soldier in the Army or a Marine. I feel that in the Army you are a soldier first and a tradesman second; in the RAF it has got to the point where we are probably tradesmen first and then servicemen. Our airmen are not trained as soldiers. They might never be put into a situation where they would be given a weapon, though they must fire once or twice to qualify. They are different.'

In case I should get the idea that the airmen were a bunch of bespectacled swots, the Corporal added that the average standard of literacy was very poor, even with people who had 'O'-levels in English. But that was just part of a national malaise – his own children were just as bad.

When I visited Swinderby, short haircuts were in vogue for ordinary civilians. This had eased one of the difficulties recruits often feel when they first arrive – the loss of practically all their hair. The barber at Swinderby, I was told, was half an ogre and half a recipient of confidences from young men grateful to see at least *one* civilian face.

The Swinderby barber, I found, had his shop just off the hangar where recruits were first lectured and drilled. He was a quietly spoken man who bore a startling resemblance to a well-known international tenor, now no longer with us. He had been cutting hair at Swinderby since 1962, at the rate of up to 180 recruits a week – about two minutes per recruit, as compared with the ten minutes he took on civilians.

In all the time he had been at Swinderby, said the barber, only *one* recruit had ever got out of the chair and run away, one sideboard removed, the other one still there. He had been sent back to have the dreaded work completed, which the barber tactfully did without comment. Reluctance was more rare now that short haircuts were de rigueur in civilian life: in the late 1960s and early 1970s it had been rather different.

Young men arriving today were more individual and easy-going than their predecessors, said the barber. They talked more about their own individual concerns and not so much about the service. 'Individual rather than collective,' said the barber in one of the rare quiet periods in his shop. 'I have to keep them tidy. They are not allowed to have a real skinhead. They don't need to have a Marine short back, sides and top – they can have it fairly long on top, provided it doesn't bulk the hat. Sometimes it is so thick and long that the hat doesn't fit properly. I can always judge.'

The barber thought that standards of personal hygiene were higher. 'All the time I have been here, I have only had one or two who have had lice in their hair. I took them out to the Warrant Officer at reception, and they were sent to sick quarters and disinfected. I put on a different gown and changed my instruments.'

Partly because they are less integrated with the men than their female counterparts at Cranwell, airwomen at Swinderby have a *slightly* more cushioned existence. 'There is a feeling that it is important for WRAF recruits to have a senior WRAF officer around,' I was told by a woman Squadron Leader, a secretarial officer whose job at Swinderby could have been filled by either a woman or a man. 'It is more so that I can act as adviser to RAF officers on WRAF problems. In general, the problems are fairly trivial. The sort of question I get from the girls is, "Where can the WRAF recruit get a pair of regulation court shoes?" A burning issue, but I hadn't the faintest idea. We give the WRAF an allowance – it is now £22 – to go and buy a pair, but the Naafi has stopped stocking court shoes. If you are an extreme size, like size eight, you can still find some at the Naafi, but not many girls are size eight. We don't dictate that they buy them from the Naafi, but we dictate the dimensions of heels and so on. Mine, I had to explain, were from the Australian Body Shop and Shoe Company in Lincoln. Sadly, I discovered, they don't do the pattern any longer. Yes, the most strange things can prove difficult.'

Women may use the same social venue as the men recruits – the Newcomers' Club, a barn of a place with all the intimacy and almost all the size of an aircraft hangar. But they are housed in barrack blocks which are separate from the men. Neither sex is allowed into the barrack blocks of the opposite sex. As the girls tend to be a year older than the men, at around nineteen, and are often thought to be more mature and possibly better educated, they tend to adapt to the (slightly more pastel-coloured) women's blocks with less hassle than many of the men.

Just before the passing-out parade of one intake at Swinderby, I went to the vintage Edwards Block for men, one of the oldest, and saw recruits getting ready for this crowning event of their hard course.

The atmosphere in this men's block was somewhat euphoric. One recruit wanting to go into the RAF police readily volunteered his main moan about his six weeks. 'The food! You don't get enough food – and you don't get enough because you don't get enough time to *finish* it! In theory you can come round for seconds, but in practice they don't leave you the time.'

In the corner of this room, which was occupied by fifteen recruits, a Corporal was asking a group of recruits: 'Have I ever lied to you?' and the group felt easy enough, now they were through the course, to hoot derisively.

A sixteen-year-old sat in a corner polishing his bayonet for the great parade. 'What I found most difficult was leaving home for the first time and getting used to the system here. I most enjoyed getting to know everyone else. I think there *is* a division between officers and airmen. We only see one another on odd occasions. Officers are a bit stuck up, really, seeing as we never see them. I think it is the wrong attitude. If we saw them more, and got to know them better, it would be better for all of us. They could come more into our social life, visit our mess.'

These venturesome comments – which I could not envisage him making until he knew he had passed – were interrupted by a recruit putting his head around the door, not seeing the officer present in time and saying, 'Anybody here got any BO-Basher? Whoops, sorry, sir!' Another round of laughter indicated that mood of satisfaction which is the keynote of the parade.

The mood was infectious. A Corporal confessed: 'I have learned more in three years at Swinderby than I did in the seven years before I came here. I am a better person for being here. I pass my experience on to the recruits, but I gain a wealth of experience through them as well. You can always learn something off youngsters. I find I can cope with things better; and that has overspilled into my family life. I have a different outlook to my own children since I came here. I can understand problems better.'

Some could not join in this end-of-term atmosphere. The woman Squadron Leader told me she had just had to do an interview with a recruit who was applying for his discharge. After seven days he had decided he did not like service life. The irony, said the Squadron Leader, was that the young man was the child of a disordered home, had a mother who lived with a series of boyfriends – and had put himself forward as a recruit because he said he *wanted* more discipline in his life.

For most, the atmosphere of parade day was distinctly upbeat. On average, for every recruit passing out in front of an inspecting officer, there are at least three family guests, who are suitably impressed by the fly-past at the precise second the inspecting officer arrives. Today there are behind-the-scenes economies even for such potent ritual. The car I had been using to get around Swinderby was suddenly withdrawn: it was the same car that had to take the inspecting officer on to the parade ground. As a taxpayer I was suitably impressed.

The fly-past, with its gusts of noise and slipstream, exorcises the reverses and humiliations of the past six weeks – the going absent without leave; the charges for having untidy lockers; and so on.

Parents and friends cluster in the Newcomers' Club afterwards, but the recruits have now ceased to be the lowest professional caste or class in the RAF – the rookies – and show it in the ease with which they fetch drinks for their parents. The successful recruits looked *more* like individuals than the awkward boys I had seen arrive.

Swinderby's Station Commander, a Group Captain, told me before I left: 'There is a danger that people will regard any institution that takes in a hundred people a week and passes out one hundred a week as a sausage machine. This is not so. They are large groups of *individuals*. Training will never work unless they are regarded in that way.'

The fall-out rate – 4½ per cent for men and 7 per cent for women – is small enough to suggest the system does work. What happens to those who stay in?

14

A Closed Circle?

'The Royal Air Force today is different in this respect. A chap *promotes himself*. It is entirely on how he performs. Once a man is in the service, we are not interested in his antecedents or where he went to school – they won't necessarily help him to be a fighter pilot or a good leader of men. I don't even think it would help him become a good Air Secretary – though it may have done thirty years ago.'

– Wing Commander at RAF Personnel Management Centre, Innsworth.

'You can say the RAF isn't what it used to be. It is getting towards facelessness. We don't have the characters around we used to. At Wattisham the other day, where younger pilots have taken over from older ones who are now flying Tornados in Germany, there was a nice touch of madness. I miss that sometimes, but getting promoted is now a serious business.'

– Squadron Leader involved in pre-promotion educational duties.

'You can't give a Tornado to a man just because his father is an Air Marshal, or you would write off a lot of Tornados. We couldn't run the organization like that.'

– Air Secretary at RAF Innsworth (Air Vice Marshal in charge of manning).

The Squadron Leader was puzzled and more than a little pained. He quite *liked* being a Squadron Leader. The trouble was that he had liked it more fifteen years ago than he liked *still* being a Squadron Leader now.

He was a slow-speaking man who weighed his words carefully. 'I have been a Squadron Leader for fifteen years, and I have no idea why I haven't been promoted. When you have been a Squadron Leader for ten years, you start to feel a certain *discomfort* – yes, I think that's the word. I have had two jobs in succession now, which haven't really challenged me to my limit, and if there's a third . . . Don't get me wrong. I've enjoyed my time in the Royal Air Force. It's been a great life. There's another thing. I have two children at public school at £1,900 a term each; the boarding-school allowance doesn't cover the lot but it's not something anyone gives up lightly. For *some* people it's one of the main reasons they stay in the RAF.'

He was an intelligent man of fifty, with many interests in civilian life, including the stock market. You would imagine him reading the *Daily Telegraph*, but you wouldn't have been surprised to find him reading the *Guardian* or the *Independent*. At the time we met he was doing an undoubtedly useful desk job; but it was plain he would have liked to be closer to aeroplanes. It was the second desk job he had been given as a Squadron Leader. For a time he had been in Intelligence, and that may have limited his future career opportunities; he just didn't know. Soon he quietly changed the subject.

The RAF operates what it calls a 'closed' system of reports which lead to promotion. I got the impression that Squadron Leader Grey (as I shall call him) would not be among its most enthusiastic supporters. Somewhere, somehow, it seemed to have left him in the dark about why he hadn't got on as fast in the RAF as he would have liked. He might have preferred a system in which *all* reports on a man are shown to the man himself; whereas in fact the RAF in practice keeps some cards close to its chest, as I was to discover.

Wing Commander Mint, a rank higher, was only forty-two. He was also very bright in his manner; pleased with life as well as himself. Promotion had come swiftly for him and, as far as he was concerned, the system could be as closed as it liked.

He left public school at seventeen and went into farming briefly before joining the RAF for a course at Cranwell College, where he completed an associated external London University degree in history, economics and war studies; then went into his first RAF appointment at RAF Stradishall as a secretarial Pilot Officer in the Accounts Flight, training young men.

'That is the point where you find senior Warrant Officers formally under you but in fact knowing it all,' he recalled. 'It teaches you a lot.'

He then went to Norway for two years as adjutant in the United Kingdom Support element at NATO, Oslo. It was while he was in the demanding and biting cold of Norway that he was promoted to Flight Lieutenant after two years as a Flying Officer, itself the lowest form of officer life except for Pilot Officers. Back in this country he was in a staff appointment at the headquarters of Strike Command near High Wycombe, looking after the organization of units. Two years later saw him at RAF Coningsby, Lincolnshire, running a clerical services flight.

His personality *and* career profile were emerging very clearly from the list of appointments: here was a man, surely, who had both earmarked himself and been earmarked by others as a fellow who should get as much varied – even if sometimes uncomfortable – experience as possible, in as short a time as possible, in order to aspire to the very highest possible rank.

At thirty he had his first 'personal' appointment, a desirable step for a sharp careerist, involving as it did learning how to deal at close quarters with superior officers, as an aide de camp to the Commander in Chief of Training Command. His next appointment was less glamorous, running building maintenance and projects at RAF Scampton, Lincolnshire; but it brought him promotion to Squadron Leader at the age of thirty-two. After two years handling the posting of ground staff at the Personnel Management Centre in Gloucestershire, he attended the main course at the RAF Staff College before moving to another personnel appointment as personal staff officer to one of the members of the Air Force Board, the Air Member for Supply and Organization. And at thirty-eight he was promoted Wing Commander at RAF Honington in Suffolk, as Officer Commanding the Administrative Wing, handling the administration of a Tornado base. He spent four months in the Falklands as Officer Commanding the Administrative Wing, and then went to the Personnel Management Centre again, dealing with forward planning.

I didn't meet Wing Commander Mint's wife, so I couldn't ask her how she saw what, in service terms, is called such a 'turbulent' professional life. But Wing Commander Mint himself was bubbling over with enthusiasm as he talked about himself and his prospects.

'I am now forty-two. Within the next year or two, who knows? . . . The average age for promotion in my branch is forty-four or forty-five. But that is changing because of a scheme whereby we look at everyone a little early so the fast runner has an opportunity to go through the field.'

With the quieter-spoken Squadron Leader Grey in mind, I asked the Wing Commander what he would feel like if promotion *didn't* arrive on, or ahead of, time. The reply was as crisp as ever: 'If I weren't promoted within two or three years I think I would begin to think the writing was on the wall and I would stay as a Wing Commander. Clearly things may not have fitted together; something would not have jelled as one hoped it would have done; and I might have to get out of the job altogether. I was a Squadron Leader for six years. After ten years I would have started to think I had been passed over.'

Plainly even enunciating such a possibility was disagreeable to Wing Commander Mint. 'I have been extremely lucky,' he enthused again. 'I have had a superb career and enjoyed it enormously. I don't want to sound like a commercial, but I have been promoted virtually at the earliest points as I have gone through. It has been a very interesting range. Diverse. Fascinating. Interesting. Lots of aeroplanes and tremendous fun . . . *Do* I sound like a commercial?'

'Yes,' I said.

At a superficial glance it seemed to me distinctly possible that the 'closed' system of promotion had in this case – and perhaps others – closed comfortably on a predictable kind of public-school whiz kid whom our grandfathers would have adjudged born to rule.

Somewhat later I met another man of the same rank, Wing Commander Glint. He was midway between the two others in his reaction to the closed system which I was to explore in more detail later: the system had served him well, but he remained of an independent and questioning turn of mind despite his own situation.

With a garden-broom moustache and sharp appraising eyes, Wing Commander Glint admitted himself to be 'very much an oddball'.

'I came in one jump ahead of National Service in 1952, without an idea of what I wanted to do,' he told me. 'I applied for entry as a radar mechanic but because of my ability – call it what you will – I was invited to go into the Intelligence side and I got through to senior NCO quite rapidly. I was serving in Berlin for six and a half years and met my wife there.'

The problem thus created may, in the long term, have been good for his career. His wife came from *East* Germany, and her parents still lived in East Berlin – a fact but for which he might still have been in Intelligence, having a more limited career.

'Intelligence and a wife from East Germany do not mix,' said Wing Commander Glint. 'It was a change of trades or it was leaving the service. I chose to become a pen-pusher. I progressed to Warrant Officer as a clerk, and was commissioned at thirty-five as a Flying Officer. I was the next one off the list for commissioning in my branch – in that respect my career pattern was reasonably typical. But I have done better than most do. Changing horses at thirty-five is relatively unusual. It is relatively unusual for someone who is a Wing Commander to reach that rank after leaving it as late as thirty-five to make the switch to commissioned officer. You really need to make the switch around the thirty mark. There have been others who have made the switch later than that and done quite well, but those who go to the very top probably switch in their early twenties.'

Still, as the son of a piano tuner and a book-keeper and pupil of an ordinary secondary school, Wing Commander Glint reckoned that the closed system of promotion had served him well. He would leave in three years' time, at fifty-five, and knew he would advance no higher in rank. But he didn't think, at one time, that he was going to get as far as he in fact had done.

And he was very keen that I should know that, however he did it, it wasn't by crawling: 'I am a fairly honest, plain-spoken man. That is not a quality universally approved. I value honesty very highly and am not prepared to compromise on it. Some people of a more sycophantic nature find it rather hard to take.'

When I met him, Wing Commander Glint was in fact helping to *run* the closed system which had earlier so beneficially promoted him – an example perhaps of the RAF's eagerness to make an individual's personal experience pay off in service terms. He was one of those at the RAF Personnel Management Centre, near Gloucester, responsible for handling the promotion of non-commissioned ranks.

The Centre contributes to the image the RAF works to create for itself as a modern, classless and united organization that calls on the discipline of the other two armed services and the flexibility and open-mindedness of civilian life.

Its very name is significant: Personnel Management Centre. So is its main headquarters across the road from the officers' mess at Innsworth in Gloucestershire: it could pass for a modern office block belonging to a computer firm determined to provide a pleasant life-style for its employees. The atmosphere is not that of an élite protecting its own, but of an open organization in which the careers of officers and non-commissioned ranks alike are *managed* in the best interests of themselves and of the service. Commissioned and non-commissioned men and women are managed virtually from the same geographical point. When I made my visit, officer promotion was handled in a separate building four miles up the road from Innsworth, at Barnwood.

Barnwood, also on the outskirts of Gloucester, was, at the time of my visit, a group of brick and concrete buildings of some ugliness and inter-war vintage; an anonymous set of government buildings turned over to the RAF. There was certainly nothing of an élitist atmosphere about the place, and I deferred my questions on the 'closed' system of reporting on candidates for promotion until I had been given a general view of the Personnel Management Centre's context and function.

There were, I was told, some 93,500 uniformed personnel in the RAF, of which about 6,000 were under training at any one time. There was an officer corps of 15,600, divided into sixteen branches, the remainder being NCO aircrew or ground trades – 78,800 of them. There were 133 people on exchange with the USA, with industrial concerns or in factories. There were about 17,000 servicemen serving overseas, of which nearly 11,000 were serving in RAF Germany, including Berlin.

'The role of this centre,' I was told at about four hundred jet-propelled words a minute by my Wing Commander guide, 'is as follows. One, the management of the careers, promotion, appointment, retirement and personal affairs of all officers and airmen. Two, the provision and deployment of all uniformed manpower, both regular and reserve, to meet service requirements and priorities. Three, advice on general personal matters and careers advice. Four, provision of computer support to the service for the maintenance of personal records and the issuing of pay.'

It sounded quite a lot to be handled from one neat IBM-ish central office block and a collection of wooden huts raised clear of the ground in a way unknown to most people alive in Britain today, except through the medium of old war movies like *The Wooden Horse*.

I said so. 'We have 390 RAF officers and airmen who will do a minimum of three years with us,' replied the Wing Commander. 'The 390 are from operational stations. We also have 560 civilians – just under a thousand people to manage 93,500 servicemen. And we keep in touch with what people are feeling on the ground. We go out and give frequent briefings about the state of the service in terms of commissioning. We go out to RAF stations. A station will get a visit from this place at least once every two years. We have just under two hundred RAF locations for a full team visit, but individuals go out to smaller centres. We work on the system of desk officers here. A desk officer for officers will handle the affairs of three to four hundred officers; a trade desk NCO will look after about 1,200 people.'

Unlike the British Army, which has a fairly simple promotion ladder for both commissioned and non-commissioned ranks, and a simpler set of 'desirable' personal attributes, the promotion structure of the RAF is of daunting complexity. But the desk officer concept is its simple keystone – there is always someone (the desk officer) an RAF man or woman can ring up about promotion or career matters generally. These desk officers sit at Innsworth and Barnwood in anonymous formations in open-plan offices; but their job can take in a substantial number of personal matters.

'I would like to be posted to Scotland because my grandmother up there is ill. Is there any chance?' Such enquiries are common, and they might be even more common were it not for a limit the RAF has set on them. The RAF, which acknowledges that its reporting system for promotions is a closed one, asserts that the relationship between desk officer and serviceman is called The Open Door. But it only operates for officers, Warrant Officers and Flight Sergeants.

Why? My Wing Commander guide said promptly: 'If we opened it up to the whole service of 93,500 people, it would swamp the system.'

This enabled me to go on to the attack about the closed reporting system. What about that? What did it mean – closed?

'The system is based on an annual confidential report – and that report is closed,' said the Wing Commander. 'The report is not read out to the candidate, although the candidate's superior officer will tell him what is written in general terms about his strengths and weaknesses. This is designed to find who is the fast runner in the system. There is a score from one to nine for individual attributes, leading to a general score of one to nine. The airman has a similar system, with assessment and grades.'

I pointed out that in the British Army, an institution which believes that men must be prepared to square off face to face in battle *and* in promotion procedures, an officer reporting on a junior officer in a way that was going to affect his promotion had to be prepared to read out any adverse comments to the man being reported on.

The Wing Commander thought a bit about that. 'Well, the Royal Navy have a virtually closed system. You have got to ensure that people report honestly on people. The Americans have an open system, and it is totally valueless.'

How so? 'You can get no subjective judgments if you are showing the person you are reporting on everything you are saying about him. You must tell him where his weaknesses lie, but not in the same words as in the report itself. A trusty servant would be offended if you said, "You are pretty thick, aren't you?" But reporting officers accept the responsibility of telling people in essence what is going on.'

There is no secret about the parcel of individual attributes on which officers are rated by their reporting officers. There are twelve of them. They are (1) leadership, (2) effective intelligence, (3) team work, (4) strength of character, (5) service and professional knowledge, (6) loyalty, (7) energy, (8) breadth of interests, (9) social attributes, (10) power of written expression, (11) power of spoken expression, (12) managerial competence.

All officers are assessed on these qualities. In addition, aircrew are also rated on the first five of these attributes specifically *in the air*.

Promotion up to the rank of Flight Lieutenant is on time served, provided the candidate gets a promotion recommendation from his superior officer. To Squadron Leader and beyond is by selection and competition. To be eligible for promotion to Squadron Leader, the candidate has to be under forty-five, to have passed a promotion examination and to have four years' seniority as a Flight Lieutenant. Candidates aged over forty-five can still be promoted, without an examination.

Each candidate of whatever rank is looked at for promotion every year by the Promotion Board (one officer from Barnwood and three from the field, all two ranks up from the substantive rank to which they are promoting, so they are not sitting in judgment on those within immediate sight of challenging them for posts). The officers sitting in judgment have the written reports of candidates thought to be fairly ripe for promotion and abbreviated career summaries on those less certain to get through. Each board member presses a green light button when he has assessed the candidate on the basis of his annual reports, but does not know in so doing whether he has actually promoted the man or not. All he knows is that he has rated the men before the board in a merit list – promotions will come off the top of that list.

Late developers? There is a proviso for those. At Barnwood a Wing Commander responsible for careers advice told me: 'A man gets looked at every year. If he is not selected for promotion for three years, we invite him here and read his tea-leaves for him. We invite him, in other words, for a full career advice interview. We can see his annual reports in broad terms, and point out his strengths and weaknesses as they are seen historically. A man can have a personality clash with a superior officer and that stands out like a sore thumb.'

The Wing Commander said that in the branches he covered, he would see about 90 to 100 men and women a year in this way. 'They may ask for an interview themselves, but they may not always get one. I would use a summary of their reports. If a man is aged thirty-two or thirty-three I would probably go right back to the beginning and say, "You made something of a shaky start on your first tour: you then came out of your shell a bit." I would think whether he was good as a professional aviator but I would note other things. In general was he good on the social scene, or did he poke off at five in the afternoon and not show any interest in the wider aspects of an officer's life?'

If a man was judged at this depth, surely it was right that he should know *exactly* what had been written about him in the past?

But this officer, too, defended the closed reporting system. 'There is not much point telling someone of forty-five or so that he is abrasive and irascible, or that he is not assertive enough – it is part of his character that is unlikely to change. I would just point out in general terms that a man's motivations had a relevance towards progress in the service. I might try to influence a chap in what steps to take. I might recommend an intermediate staff studies course, which means an eighteen-month correspondence course conducted by the RAF Staff College at Bracknell in Berkshire. This is certainly a prerequisite for going to Staff College, where the high-flyers go. If a chap shows no interest, he cannot be surprised if his masters don't see his motivation towards a full career in the RAF.'

Unlike the Royal Navy, which boots out officers if they are not thought fit for promotion beyond a certain point, the RAF appears to concede that there are some men, conscientious and able servants of the service, who simply do not want to play the career snakes-and-ladders game by chopping and changing jobs to get the widest experience and generally buzzing about to get the highest number of steps up the ladder.

Staff at Barnwood told me there were a significant

number of people who were not seeking to be high-flyers – they were perfectly happy to be professional pilots or navigators. And that was considered a perfectly respectable aim. If a man had not been promoted to Squadron Leader (which meant he was guaranteed a job until age fifty-five) by the age of thirty-six, he was looked at to see whether he could be offered assimilation into a full career as what was called specialist aircrew.

'Second-rate citizens? They are looked upon as the backbone of the flying branch,' one officer told me. 'They will continue to fly in junior officer flying appointments. At thirty-eight they have become specialist aircrew who still compete for full career promotion until they are forty-five; if not promoted to Squadron Leader by that age, they will remain flying as specialists with limited promotion prospects.'

Such possibilities could be particularly attractive to late developers or those who join the RAF later than most. All right, I was told, they were not going to become Air Marshals, but they could be jolly good Squadron Leaders and perhaps Wing Commanders.

But what of the fact that some people were presented to the board in full parade, as it were, as likely promotion material, while others were presented only in profiles of their performance and promotion prospects, which suggested in advance that they were thought inferior material? Was that strictly fair?

Board members think it *is* fair. One of them told me that, in the general duties branch, there would be far too many candidates for promotion for the board members to consider them all in sufficient detail. So there was a pre-boarding system. A Wing Commander and a Squadron Leader spent about three months going through every confidential report of about 1,000 candidates, and classifying the men into three streams: A (likely), B (not quite ready yet) or C (unlikely).

'But every one is presented on paper to the promotions board, though some are in profile form,' insisted a Wing

Commander involved in this pre-boarding work. 'The president of the board has the actual written reports; the other three have screened fiches and books of profiles. They consider the reports *without discussion with one another*, and then they are ready to give someone a mark from nought to nine. We write down the aggregate: therefore they don't know whether they have promoted him, they only know the score.'

An Air Board considering promotion from Flight Lieutenant to Squadron Leader can take four weeks to examine 1,600 people – 1,000 from general duties, air; 600 from air traffic control, fighter control and so on.

But there will be only 130 slots for Squadron Leaders. And, by the time they have picked the top 125 scores, it may be obvious to the president of the board that there are ten people with the same marks competing for those ten Squadron Leader promotions. The board looks through the ten men again and puts them in order of preference. How that is done is left entirely to the judgment of the board itself. If a candidate is particularly young, they may think he can afford to wait a year. If the candidate is much older, they may reason that if he does get promotion now, he will not go much further, and make him number seven on the list. Then all the board jointly do a mini-boarding to decide on the final order of preference. Those below the acceptance line will be put on a reserve list which may operate if anything goes wrong with the chosen ones; but if they are not picked up at the time, from that reserves list, they 'go back in the hat' next year, without any special priority.

The first sixty-five on the list will be promoted on the July list, the second sixty-five on the January one. If the people fall out from the July list, it is men from the January list who are brought forward to replace them, not the reserves. This is thought to be a way of keeping up standards to the highest possible level.

In civilian life it is not rare for a poor professional reputation to be self-perpetuating. A man is thought bound to be poor *this* time because he was poor *last* time; and this fact may tell repeatedly against him. The RAF likes to think it has budgeted for this destructive human idiosyncrasy.

If a man has been presented five times running as a B (not quite ready yet) or C (unlikely) candidate, the board is required to look in detail at his past reports to confirm that his potential has not been missed by some oversight. Probably this board would go right back to his very first reports to see if they can detect anything that would explain the candidate's apparent weakness.

There is a further factor which can minimize past errors of judgment. People remain eligible to be considered by a promotion board until they are three months away from the end of their service. The board might tell a candidate that he would have to agree to stay on in the service until he was fifty-five or do a minimum of three years in his new rank; but he would still stand a chance of promotion. There is only one proviso: a man would not be promoted merely because he was a good worker in his present rank and the service wanted to keep him for that reason.

Airmen, at the time of my visit, were boarded at Innsworth itself, where eventually it is planned the officers' boarding will take place as well. Though there is arguably much more scope in the RAF than in the other two armed services for jumping the professional and social barrier between officers and other ranks, there is some logic in a partial geographical separation. Intentionally or not, to some observers it may seem to mirror a difference in how officers and airmen are assessed.

The Wing Commander in charge of airmen's boarding, a man who had crossed that chasm himself, put it this way: 'An airman comes in and is given basic training. In the course of his career he will be able to build on that by in-service training. Qualities that are looked for in officers are much more wide-ranging than for airmen, so the scope of annual reports for airmen is much more concerned with the day-to-day mechanics of everyday work. I don't want to sound snobbish about this, as an ex-airman myself, but it concerns itself with the basic elements of doing the job.

Men doing that job are demonstrating the personal qualities necessary within that environment.'

The structure of airmen's professional life rests on two lists. List One is devoted to technical or highly specialized men whose promotion is determined to a great extent by their technical competence rather than by larger aspects of personal character. List Two is devoted to the support trades: not those directly responsible for keeping the aircraft flying, but those enabling others to do so, from clerks to policemen.

List One men advance on time of service and technical accomplishment. To List Two there is a different approach, merit being assessed under a number of headings – thirteen of them, each to be marked on a scale of one mark to nine. Very few men, in the result, are assessed under five marks, and experienced members of the promotion board say that there are problems of judgment when a whole lot of men get around seven – which is by no means unusual.

Like the officer, the airman is assessed once a year in a confidential report. But unlike the officer, he is judged for promotion by a board consisting of only three, not four, members – and these are all officers, though an airman of superior rank may contribute to reports. The minimum rank on a promotion board is Flight Lieutenant. The board president is from inside the Centre, the two members from outside. For Flight Sergeants wanting to be Warrant Officers, the board is always headed by a Centre Group Captain.

'Aircraftmen candidates for promotion to Corporal,' said the Wing Commander in charge, 'are now listed according to performance on computer rolls, scrutinized by my staff. The recommendation to select or not is made by a Squadron Leader and is vetted by me and by my Group Captain before being passed to the Director of Personnel Management (Airmen) for approval. The Director is an Air Commodore. Sometimes we think the board gets it wrong, but the proportion of changes is very small indeed. Probably more than 99 per cent of selections are approved.'

But the Wing Commander has to be a hard man if necessary. The one I was talking to said he personally believed that, if a man had committed a crime in the last two years, and hadn't done something considerable since to rehabilitate himself, then two years were too short a period to consider him for promotion.

But there were, he said, even harder decisions that had sometimes to be taken for the good of the service. 'Equity of treatment is something we concentrate on. Seen from station or unit level, a man may or may not have a case. When you look at the global level, it assumes different proportions.'

I didn't take the point, and said so. 'Well, if a man becomes sick or is in a road traffic accident he is not fit to fight; he is not fit for worldwide service and therefore his promotion prospects are inhibited. Perhaps after being ill he is given a support job at his station, but you can't then turn round and say that because he is performing well within those limits, he should be promoted. I therefore have a harsh decision to take when I recommend that he should *not* be promoted.'

Illness or other misfortune may also affect a man's chances of having his contract extended. The Wing Commander said the RAF took the view that while, as good employers, they would do all they could to honour a current contract, they would not inhibit management flexibility by extending the service of such a man.

'That is the kind of difficult decision I take day by day,' said the Wing Commander. 'Say a man's wife has left him with two kids. We place him next door to home so the problem can be contained. But should we extend his service, when we know we will want to move him? It is debatable whether he can fulfil his full service obligations when his conditions are as they are. Sometimes the interests of the service get pushed into the background if you're not careful.'

Not, I guessed, when the Wing Commander was around, though such cases can be referred to the Compassionate Appeals Tribunal.

If twelve Warrant Officers are needed, about three times as many Flight Sergeants will be reviewed. The board consider each individual from the list of names produced by the computer and grade him or her according to merit. An individual can be high in numerical score order from the computer listing but not be selected.

It is admitted that this process is very subjective, but the Wing Commander I was speaking to had devised his own system of marking, which he considered made things as precise as possible. When he and the other two boarding officers sat down in his office with a pile of annual reports three feet high, he said, he required them to give marks out of one hundred – unlike some other board presidents who gave marks out of fifty, but with each mark liable to be given a minus or plus after it. Such prevarication was obviously not for the forthright Wing Commander, who took the view that a mark was a mark was a mark, and that was *it*.

The airman promotion results then go up to a Group Captain and thence to the Air Commodore who is head of the management of airmen, who has the final say. And, officers pointed out to me, the nicety of it was that the results were then sent to the drafters – a completely different board – to decide who should be posted to which post. There could be no question of fiddling a man into a specific job by promoting him, or depriving him of it by not promoting him.

The process of changing horses from airman to officer is also a democratizing factor. It is commonplace in the RAF for men to ask their commanding officers for commissioning. If approved, they present themselves at Biggin Hill Officer Selection Centre to be selected on the same basis as the other candidates – except, perhaps, that their knowledge of the service may give them a favourable edge. It is almost equally common for a good airman to have it suggested to him by his superiors that he should apply to be considered as an officer – a procedure the British Army might look at askance. If he applies to Biggin Hill and is successful, he goes to Cranwell for officer training just like a new entrant presenting himself, from the first, as a would-be officer.

The gap between airman and officer, it is claimed, is easy to bridge if the innate ability is there. No doubt, I thought; but some arbitrary judgments are having to be taken, in this system, about what *constitutes* ability.

Before I left Innsworth, I saw the Air Secretary himself – the Air Vice Marshal in charge of RAF manning. He was a large, softly spoken man. He was armed with two large folders, one of them buff, the other green. 'There is nothing arcane about our promotion system,' he said, opening up the files on one candidate, whose name had been carefully erased. In the confidential reports file, colloquially called a 'jacket', there was a report on a man who had been approved on one page for immediate promotion while on another was written off as 'of no further potential' – in other words he would get no promotion after *this* one.

The two judgments, I said, seemed to be in conflict. Quite, said the Air Vice Marshal, which was why he had taken a look at them. Further digging in the reports had revealed that the man was conscientious, with a good dry sense of humour, but a hard taskmaster and subject to tunnel vision when under pressure – things which had not blocked his path to immediate promotion, but *would* affect his chances of going higher. An apparently arbitrary opinion had been well based after all, it seemed.

The Air Vice Marshal also waved the two files in front of me to emphasize the fact there were *two* of them. One was the confidential reports 'jacket', the other was the man's personal file. The personal file, he said, was kept for the purposes of pay records and the like. It contained information about schooling, family background, father's occupation and so on, together with any illnesses suffered.

'This personal file will never be made available to the board,' emphasized the Air Vice Marshal. 'You are promoted purely and simply on what you *do*. Any seepage of any other opinion is bureaucratically prevented by separating the two pieces of paper – the personal file and the confidential report jacket.'

It is just as well there are rigid safeguards against fiddling favourites into the top positions. Promotion in the RAF has taken on a greatly increased seriousness in the past few years. The Air Vice Marshal said he found this seriousness to be the main change in the philosophy of promotion since he had joined the RAF in the days of National Service.

He put it like this. 'In my lifetime in the service, personal aspirations have perhaps become more sharply defined. I think there is a much greater tendency for a man to ask where he is going to be in ten years' time. One can point in the RAF to a series of facts and occurrences which have militated towards that changed attitude. When I joined, I could expect to serve in many different countries of the world in which, generally, my standard of living would be higher than it would be as a serviceman in the United Kingdom. There was a fair amount of rough and a fair amount of smooth as well. These opportunities are largely gone now.'

The result, said the Air Secretary, was that if you were seeking for the RAF to improve the quality of your life, generally speaking the *only* way you could do it was by promotion. If you added to that the introduction of the military 'salary' around 1970, when a lot of material benefits were transformed into taxable cash, it was easy to see why there was now more focussing on the rank and pay structure: the fringe benefits had almost all gone. Now there was no provision of domestic assistance, for instance; the only way you could improve your domestic situation was by getting cash, and this was tied to rank.

Wasn't all that rather Soviet? 'Any comparison between this Air Force and the Soviet Union – no way, no!' thundered the Air Vice Marshal. 'A fighting service cannot depend on time-servers. I have two management principles. One, I am a very arrogant man who doesn't like to be told he is wrong by a subordinate. There is only one thing I like less which is (two) to find out I was wrong and my subordinate didn't tell me. Most RAF officers operate in that way.'

But the Air Vice Marshal did concede that the sharper focussing on rank might, if you weren't careful, lead to men becoming more promotion-hungry than promotion-worthy. Every so often, he said, someone would get a comment on his confidential report: 'His attitude to promotion is tending to dominate his attitude to other things.' That would temper a lot of otherwise good work the man was doing. Officers who smarmed up to their superiors could get their come-uppance, too, in comments like, 'I am suspicious of this man's sycophancy.'

I laughed and said one would grow very old indeed before one encountered such a comment in a civilian organization; and the Air Vice Marshal and I parted good friends with his cordial parting shots ringing in my ears: 'On the Air Force Board (the top governing body) at present we have an Oxford graduate, a London University graduate and two products of London schools. Two members of the board are former National Servicemen. I don't think there is any "standard Establishment" pattern there. And we are proud of that. Remember we are a meritocracy.'

It is a good and confident assertion with which to confront a late-twentieth-century civilian public.

PART IV

PUBLIC CONTACTS

15

PILOTS AT SIXTEEN

'The air is the hardest taskmaster of all, and it is a three-dimensional one. Air Cadets, aged thirteen and upwards, have to be alert and they have to learn new skills fast. Those who make it are conscious of it. They walk ten feet tall.'

– Squadron Leader in charge of 618 Volunteer Gliding School, West Malling, Kent.

'Mishaps? We have the odd one. I think the mishaps here are the occasional heavy landing. In the past year we have had no accident at all.'

– Training and Supernumerary Officer of 618 Volunteer Gliding School (working in a stockbroker's office in civilian life).

At over a thousand feet above West Malling airfield in Kent the glider pilot instructor told me that, in terms of sheer nerve, there weren't two classes of cadet pupil.

'The Air Training Corps boys have more athletic ability and the Combined Cadet Force guys are more academic, but it evens itself out because in nerve they are about the same. Sometimes we get people who are very bright, but they have no sporting ability and they tend to take a bit longer to go solo for the first time. But on the other hand, you have to have intelligence to understand what is going on. No, there are not two social classes with the Air Cadets.'

For the Combined Cadet Force boys from the grammar and public schools and the Air Training Corps boys from a broader band of educational institutions, the very first taste of flying is this silent soaring at 1,000 and 1,500 feet above the British countryside. It is an emotive and exhilarating experience, watching the fields recede as the steel winch rope pulls you faster and faster, higher and higher, feeling the nose tilt disconcertingly down and then the cable disconnecting at the peak of the climb, and then swooping free in pursuit of upward thermal air currents from roads and other heat collectors in order to stay aloft as long as possible.

For some cadets, it is the single most exhilarating experience of their lives so far. One eighteen-year-old who had just completed his first solo flight said on landing: 'What you feel is hard to explain. You feel suddenly totally in control, even though for the past two or three flights the instructor has been letting you handle it by yourself anyway. Nothing at all in my life has compared with it, even though you are thinking too much about what you are doing to feel much. Once you get to the top of that launch and release the cable, you can sit back for a few seconds before doing anything else, and those moments are very exhilarating indeed.'

That cadet was working as a technical clerk for an electronics firm, but hoped later to return to the gliding school to become a Staff Cadet and help others to learn to fly.

Some cadets do their own mental contortions to remain calm on their first solo. A sixteen-year-old said after stepping back to earth: 'I felt that there was still an instructor in the back of the glider. For the last few flights you have had, your instructor has been so silent and left it to you that you can well imagine on your first solo that he is still there. I was not *that* nervous, which surprised me slightly. I can only think that I had it at the back of my mind that there was still somebody sitting back there if something went wrong.'

Sometimes – fortunately rarely – something *does* go wrong. When a twenty-year-old cadet, son of Pakistani parents, was on his last flight before he was due to take his first solo, he turned to land and found that a Land Rover was being driven across the runway. The instructor had to resume control to miss it. The civilian helper who had been driving the Land Rover was invited never to show his face on the airfield again, and the twenty-year-old's first solo was postponed.

'It shook me up a bit, that incident,' he told me later. 'It was probably just as well my solo was postponed, though I had done forty-two launches by then. I continued gliding for the rest of that day, which was a Saturday, and first thing on Sunday morning I went straight out solo and was successful. I was not too nervous: I thought of all the hard training I had had from the five pilots who had taught me.'

At the time I spoke to him he had just been promoted to Cadet Warrant Officer in the Air Training Corps, and had asked to return to West Malling as a Staff Cadet when he had got his college examinations out of the way. Eventually, he told me, he wanted to be a pilot in the RAF.

What all these three cadets who had just flown solo for the first time had in common was a remembrance of a formative experience and a certain sadness that, with the first solo, their glider flying days were over unless they were coming back as Staff Cadets to help others fly: the RAF is willing to teach young men to fly but not to provide them with free private flying thereafter. Future glider flying has to be earned by putting something back in terms of service.

For a very high proportion of men who eventually reach the Royal Air Force through either the public schools or the comprehensives, gliding at schools like West Malling is their very first experience of flying – perhaps slightly more valued by the comprehensive boys of the 39,000-strong Air Training Corps than the boys of the much smaller air section of the Combined Cadet Force, since the fathers of the latter may spend more time jetting round the world on business or professional trips.

An Air Cadet – the equalizing term by which both

Combined Cadet Force and Air Training Corps boys are known – can (if bright and persistent enough) qualify for his first solo flight in a glider when still only sixteen. The high-performance Vanguard glider in which I was strapped, ahead of the pilot, cost £25,000 and was one of seven available to 618 Volunteer Gliding School at West Malling. They were soon to be replaced with eight even more expensive Viking gliders. In few areas can a sixteen-year-old be entrusted with such a sizeable investment; but the RAF itself funds such schools as an investment in its future while knowing that the sixteen- to nineteen-year-old young man (or woman in some squadrons today) may never join the RAF. The RAF hopes that at least the pupils will have become more interested and disciplined citizens.

I found the forty-foot wingspan winch-launched Vanguard glider comfortable but spartan, the dual controls moving against hands and feet in a way that eventually allows the pupil to take over from the instructor. As a pupil would be, I was told how to strap myself in and fasten the cockpit cover, how *not* to jettison it by pressing the wrong button and how *not* to obstruct the air brake lever when coming in to land.

The chief instructor, with his black moustache and dark glasses, looked every inch an RAF man. In fact he had never been in the RAF – indeed, had flunked the aptitude test when he had applied at Biggin Hill to join the RAF. It is a useful irony that the RAF now manages to make use of the frustrated skills of would-be RAF pilots by channelling them into initiating youngsters into their first, perhaps crucial, experience of flying and its demands. Only a third of the instructors here have been in the RAF. A surveyor by profession, he had been an instructor for seventeen years.

'There is no lift around today,' apologized the chief instructor as we circled over West Malling air station – famous as a badly bombed Battle of Britain station in the 1940s, and now the perhaps precarious home of an air engineering company and the gliding school. 'But we have sometimes had as much as six minutes in the air from a thousand feet.

On average a boy has forty-five launches before going solo. Where the instruction periods are concentrated, it can be as little as twenty.'

In gliding, conversation is much easier than it is in any sort of powered flight. What qualities, I asked, were vital in learning to fly? Not necessarily intelligence, I was told, but certainly good sporting prowess, because that was good for co-ordination and judgment: 'Even a simple thing like playing table tennis or riding a bike helps.'

It didn't *seem* we had been aloft for six minutes when we zoomed down on to the grass of the airfield, with another glider, below and to the left of us, coming down on the airstrip itself; but we had been. It is easy to see how gliding can become addictive, but I was sternly told that West Malling was in no sense a mere *club*: 'It is a reserve unit organized on RAF organization lines, administered by the RAF and staffed and manned by Air Force Reserve officers. That applies to the twenty-eight schools around the country.'

My meticulous informant was the Squadron Leader (retired) in charge of the school. His civilian job was being a quality executive with a large electronics company. Those who worked under him at West Malling, some fifty people, included policemen, solicitors, bankers and a British Petroleum executive.

In many ways West Malling is typical of the gliding schools. Each has an adjutant for administration, a technical officer, a chief flying instructor responsible for flying training, a deputy chief flying instructor, a supply officer responsible for stores and a training officer responsible for calling and promoting cadets for training.

In many ways it is *not* typical: it is in fact the largest of the sixteen gliding schools which were still using winched gliders at the time of my visit. And, unlike most schools, West Malling is not based on an RAF station, where tenancy would be fairly secure. The land is owned by Kent County Council, and the lease *could* be terminated by the council: perhaps in accordance with a political whim. There are several schools in such a situation. A gliding school at Tangmere, Sussex, closed in 1975 and has not been successful in finding a new home in the south-east – since, unlike Army cadets, who can train in any old local wood, the Air Cadets need great amounts of space.

'We are worried about the future,' said the bulky and fatherly Squadron Leader. 'Although we enjoy a six-year lease, it could be the last one we get. You are very much at the mercy of the political mood of the day.'

It is not a doubt with which a twenty-five-year-old institution like 618 Volunteer Gliding School should have to live. But such are the facts of life. The instructors are adept at ensuring that such facts do not cloud the enthusiasm of pupils; they are a highly disciplined and trained force, though most are inveterate civilians.

All instructors begin as untrained instructors and are taught to fly. As passenger-carrying pilots they are allowed to take cadets for air experience but they are *not* allowed to teach them to fly. They then go on to an instructors' course at RAF Syerston, home of the Air Cadet Central Gliding School, near Nottingham, and the Headquarters for Air Cadets at nearby RAF Newton. If successful, they are given what is called a C category, which enables them to give instruction to cadets. But even now the bureaucracy will not allow them to send cadets solo; and they have to be checked by a chief flying instructor every three months. The next step forward is becoming a B instructor, which enables the teacher to do two more things – to act as duty instructor of the day and to send someone solo who has already been sent solo. Only in the A2 category is he permitted to send boys solo for the *first* time. The most senior an instructor can get is to become an A1 instructor. There was only one at West Malling at the time of my visit – the chief flying instructor who flew me.

Such conscientious grading is considered necessary when dealing with matters of possible life or death. Both A Flight, which operates on Saturdays and comes from the ATC, and B Flight, which meets on Sundays and is composed of the CCF pupils, are obliged to attend a minimum of three sessions a month; and the school operates every weekend of the year and on every bank holiday except Christmas Day. There are also nine-day intensive courses, one at Easter, one during the Spring bank holiday, three in the school summer holidays and one at Christmas from Boxing Day.

It is a human production line which embraces between 10,000 and 12,000 flights a year and trains up to solo standard some 115 pupils a year, about half from the Air Training Corps.

I did not spot any non-white faces during my visit, but I was assured they existed. 'We certainly do have people from the ethnic minority groups,' said the Squadron Leader. 'If you ask me how well they do, I would have to be careful. Frankly, we find sometimes, because of the language barrier, it can be difficult. On balance, we probably find we have to work a little harder with them. But I would say that 15 per cent of our intake are from ethnic minorities, and that is quite good in relation to the total populations. We have some who are oriental, Caribbean, Asian and African.'

Since West Malling handles about 20 per cent of the solo training carried out by all Britain's sixteen gliding schools, the atmosphere is one of high pressure. The motto on the crest, *Valamus Acriter* (We Fly Harder), is a fact, not a pious hope; it has to be. The ground facilities would not be accepted happily by a civilian flying club. There is a cadets' dormitory with old wooden cabinets, iron camp beds crammed close to one another and primitive lighting. 'Shades of the RAF of many, many years ago!' said the officer who showed it to me. In the cadets' dining room, about the size of a large living room, the spartan institutional drabness was relieved only by a space invaders game machine. Even in the instructors' lounge the atmosphere was puritanical, relieved only by a suspended model of an old Kirby Cadet Mark 3 glider and a small bar (rarely open). The Squadron Leader had a fold-up camp bed in his office, where he often uses it. 'It keeps you young,' he said.

The fact that it is flying life on a shoe-string probably adds to the excitement rather than lessening it.

One sixteen-year-old from Mid-Kent College, where he was doing an engineering course, was intensely disappointed because turbulence caused the postponement of what would have been his first solo: 'I have always been interested in flying, right from the age of eight or nine. I would like to get into the RAF as aircrew. I would like to be a pilot, but you need A-levels for that and I am not going in for A-levels; I have always been interested in practical things. My college is mainly to do with stuff like cooking, hairdressing, bricklaying, carpentry – very practical sort of stuff. So I will probably become an air loadmaster, I hope.'

His frustration at the weather was not unusual: it is one of the most irritating aspects of glider flying. This sixteen-year-old had been on thirty launches, but ten of them were marked down to 'familiarization', which means getting used to the controls again after periods of non-flying. He complained that he didn't think he was learning fast enough: 'It is frustrating when you turn out at the weekends and there is bad weather, but there is nothing you can do about it.'

A lot of pupils find such frustrations so minor that, having qualified, they return to the gliding school as training officers on the RAF Reserve. I met one twenty-one-year-old from a public school near London who said that his civilian job was arranging settlements in a stockbroker's office, and that ultimately he wanted to pursue a career on the Stock Exchange. He said his present job was well paid, and that if he went into the Stock Exchange, it would be better paid.

Then why was he spending much of his time teaching teenagers to fly, miles from his own home? 'It is just different, really. It is a complete contrast to being stuck in an office. You feel you are achieving something at the end of it. Stockbroking is a fast-moving world. You have got so much work to do that you feel you are taking one step forwards and ten steps backwards some of the time. But if I take a student for his first ride and from there he goes on to do his first solo, at the end of it, you think, "I taught him that."'

The son of a petrochemical engineer, he had joined the Air Cadets when he was thirteen and had flown his first solo – at West Malling – when he was sixteen. I asked him, too, how different he found the public and grammar schoolboys of the Combined Cadet Force from the comprehensive schoolboys of the Air Training Corps. Generally, he said, the Air Training Corps boys were more enthusiastic, partly because a lot of them joined up in the Corps and came to the gliding school as an alternative to doing community service.

'You don't get ATC cadets with long hair and things like that,' he said. 'Who makes the best pilot? It is difficult to say. Normally everyone goes solo, but the Combined Cadet Force would have the edge. That is not to say there aren't Air Training Corps cadets who are equally or more intelligent, because there are.'

Other, more experienced, instructors said they were often surprised at how often apparently not very bright Air Cadets went successfully through the course, after a very short period of tuition and without the experience of even driving a car. A fifty-seven-year-old Wing Commander (retired), now working as a civil servant for the RAF, arranging flying scholarships or sixth-form scholarships, said he found teaching boys to fly a natural extension of his bureaucratic work and something that gave him a great deal of stimulation and satisfaction. The supply officer turned out to be a tutor at the Royal College of Art. He saw no disparity between the two activities. There was a physical side to flying and an artistic side to his main job, he said, and they went well together: he needed the one to recharge his energy for the other. He had done his National Service in the RAF, and found himself for two years with an ATC gliding school. 'They were rather happy days and I didn't want to go out at the end of the two years. I am glad to be still taking part in gliding and teaching.'

At West Malling I came across only one girl. She was an eighteen-year-old Flight Staff Cadet who joined because she 'liked the atmosphere – it struck me as very friendly'. She wanted to fly, but her aim was to become involved in the administration of medicine. She had long since flown solo, at RAF Cosford, in a motorized Venture glider, but she told me thoughtfully that she had mixed feelings about not being able to become a pilot in the RAF because she was a woman.

Her reasoning was an illustration of how the RAF is able to attract, and keep within its orbit, young people who are by no means vapidly gung-ho in their approach to life.

'I am not sure I would have liked to be able to become a pilot, because of the moral aspect of *bombing* people. But if I were a man I would consider being a *fighter* pilot, because they are defending us. But anyway I want to do medicine, and I think I would prefer that, even if I were a man.'

This young lady seemed very much at ease, helping to record the launchings and returns of the flights of the day. But I asked the Squadron Leader in charge what was the position of women in the ATC and CCF at the moment. He told me an increasing number of ATC squadrons in Britain were being authorized to take female cadets. West Malling had trained a number of female pupils and girls as staff cadets, knowing that if women wanted to become fully blown air traffic controllers in civilian control towers they must have a pilot's licence, so that their employers knew *they* knew the problems of the pilot.

Generally speaking, the first taste of flying young people can receive is in a man's world, a world in which flying is almost a 'religion'; and those who have had to move away from it often return to it as 'priests'.

The duty instructor and adjutant at West Malling was a forty-five-year-old self-employed electrician in civilian life. He joined Leeds University Air Squadron, was trained as a pilot by the RAF while still at university and had not been able to fly since – at least, not until he heard about the gliding school near his home in Kent. Then his head, which told him he had to earn his living, and his heart, which wanted to be airborne, at last came together, and he began to be a really happy man.

'And you can go on until you are fifty-five,' he said in the tones of a man who had found his spiritual home.

In another aspect of Royal Air Force life that closely touches the civilian world, such an age would *not* be a recommendation.

16

THE RESCUERS

'We in the Royal Air Force Search and Rescue Service, serving civilians as well as the services, are the acceptable face of defence spending forty years after the last European war. For me, one of the most enjoyable things is the liaison with the other rescue services, and with the public.'

– Pilot of RAF Search and Rescue Wessex helicopter, a Flight Lieutenant of E Flight, 22 Squadron, based at RAF Manston, near Ramsgate.

'We serve West and East alike if they are in trouble. We take off injured Russian trawlermen and put them in United Kingdom hospitals. The Russians and Poles fish near our shores and it is a well-trodden route – no hassle. If it is one of the Russian boats you tend to think there is a KGB man aboard, but there probably isn't – just fishermen doing their job.'

– Flight Commander of 22 Squadron Search and Rescue Flight, RAF Manston.

From this angle, the White Cliffs of Dover were not as reassuring as usual.

Politically, the Royal Air Force Search and Rescue Service –which helps threatened service personnel and civilians alike – may indeed, to the general public, be the acceptable face of defence expenditure, a social service conducted with military precision and discipline and, as such, a great public reassurance.

But its tasks are always demanding and sometimes hair-raising – as when we hovered in a Wessex helicopter alongside the high Abbot's Cliff, directly above someone who had toppled over the cliff and got stuck on a crevice fifty feet down. The swirling rotor blades were within fifteen feet of the cliff face – the nearest a Wessex is allowed to go – and a sudden gust of wind could (if it caught the pilot unprepared) ram the blades into the side of the chalky cliff, sending the helicopter and its occupants, including me, down four hundred feet to the rocks below. A down-draught, to which helicopters are notoriously prone, could also send us lethally on the straight-down route.

Fortunately, this particular incident was only a crew training exercise. The trapped 'person' was in fact a flare, which was shot into the cliff by the navigator (an officer) and retrieved by the winchman (a non-commissioned officer) who was dangling at the end of a steel cable that can be extended to three hundred feet and put the swinging winchman in considerable danger. The only passenger aboard the Wessex during this particular twenty-four-hour spell of duty by E Flight of 22 Squadron from RAF Manston, three miles from Ramsgate on the Kent coast, was me – professionally kitted out with flying overalls and headphoned helmet but otherwise all too aware of my limitations, as most civilians would be in such a situation.

At a moment's notice, the Wessex could have been diverted to a real emergency, and I was warned it was quite likely that it would be. Training goes on all the time: later a 'body' was to be recovered from the sea, and the passengers of a cross-Channel ferry had a surprise when we put the winchman down on to a hastily cleared rear deck. But the possibility of a real emergency intervening is quite concrete, especially during the leisure boating of the summer season and the gales of winter.

Already that month, there had been ten incidents in which lives were, or could have been, at risk. An overdue fishing boat had to be investigated. An aircraft lost its radio contact as it came over the coast and had to be assisted. A search for a swimmer, said to be in difficulties off the Isle of Sheppey, revealed that the swimmer had not in fact been drowning, only waving. A German yacht put out a distress signal but was helped by other ships. A motor cruiser went aground north of Margate, and E Flight provided the air cover, hovering over the stricken ship until the lifeboat arrived. There were no less than five heart attacks on which E Flight had to take action. One man with a heart condition was taken by Ramsgate lifeboat to the shore, where a Wessex lifted him to Margate Hospital. A man had a heart attack on a Sally Line ferry, *Viking*, and was lifted off and taken to hospital at Canterbury. A woman had a heart attack on *Viking II*. Another man had a heart attack on a barge at West Mersea; a doctor was deposited aboard the lifeboat as it raced to the scene, but found the man dead. A crew member of a Polish tanker had a heart attack and was found to be dead before a Wessex could get to the ship.

These were merely the incidents which had happened to one flight of one squadron of the RAF Search and Rescue Service. All over the country, similar incidents were happening, with either Wessex or Sea King helicopters as, virtually, National Health Service ambulances in the case of heart attacks where minutes, even seconds, may be vital. The four other flights of 22 Squadron at the time of my visit to Manston were at Leuchars (Scotland), Leconfield (north of Hull), Chivenor (Devon) and Valley (Anglesey); the headquarters being at Finningley (Yorkshire). There was another squadron, 202, which flew Sea King helicopters based at Boulmer (Northumberland), Coltishall (Norfolk), Brawdy (South Wales) and Lossiemouth (Scotland), all with (as in the case of E Flight, 22 Squadron) two aircraft to a flight and four sets of aircrew.

Each flight has to contend with about one hundred and twenty scrambles a year, two-thirds of them needing positive action on arrival.

During one fairly typical year, 1986, the RAF and the Royal Navy together had 1,398 call-outs, for which the RAF was responsible for 992. The RAF rescued 638 people compared with the Royal Navy's 159, some 90 per cent of them civilians. This was done with a force of ten Wessex helicopters in 22 Squadron, eight Sea Kings in 202 Squadron and one fixed-wing Nimrod on one-hour call in either Scotland or Cornwall. The RAF mountain rescue teams saved seventy-three people.

It is no wonder that experienced RAF men say that the engine noise of the characteristic yellow helicopters reminds a sometimes touchy public that RAF aircraft noise is always connected with their protection – and most *directly* so in the case of the Search and Rescue Service.

RAF Manston has an enormously broad landing strip, a legacy of the Second World War, when shot-up aircraft returning home needed the maximum width to wobble on to. In all other respects it is a Lilliputian RAF station. Its officers' mess is the size of a small country house, with a restaurant seating only sixteen. It has a very thin administrative spine, and by the time you reach those 'lodgers' at RAF Manston, E Flight of 22 Squadron, housed in a self-contained bungalow near the helicopter pad, you are in a very lean situation indeed. When one aircrew is flying, one other is on bleeper stand-by, and the other is off duty. Telephone calls to the main flight switchboard have to be answered by ground crew – chiefly electricians or engineers. The fact that the system usually operates smoothly is plainly due more to the dedication of men who are *always* on active service, even in peacetime, than to the British taxpayers who pay leanly for this lean operation.

Each flight arranges its duty hours to suit its particular needs. Areas containing a lot of mountains are obviously different from those with a lot of shipping. I was with E Flight of 22 Squadron when its day crew came on – at ten in the morning – mingling with the night crew, who were in the process of being debriefed. This is a valued part of the day. Without it each three-man crew (pilot, navigator and winchman) would mix only with themselves in the monastic conditions, making their own meals in the tiny flight building and resting on beds in their rooms in the same building. As it is, it is still a very small, often supercharged, world.

The Squadron Leader, who spends half his time going about the country visiting the spread-out flights of 22 Squadron by car, and who was at RAF Manston when I visited it, is greeted with more enthusiasm than usually accorded to enquiring senior officers. He represents contact with the bigger outside world. He is the only man in the Search and Rescue Service who is *never* called by his Christian name by junior officers or non-commissioned officers, unless he specifically requests it. Everyone else, I saw, addressed each other with a matiness I had not encountered elsewhere in the RAF – though instantly getting on surname or 'sir' terms again when in the public eye.

At ten in the morning, the three men who formed the oncoming crew had a briefing which covered everything they needed to know. The Flight Commander, a thirty-eight-year-old Flight Lieutenant, was tall, quietly spoken and given to large cigars: a not untypical pilot. The navigator was a thirty-four-year-old, aggressively genial and heavily moustached Flight Lieutenant who thought the demands of the service had helped him overcome the drink problem of his even more aggressive youth. The winchman, the only non-commissioned officer in the crew, was a bearded Welshman. He was a master air loadmaster by trade and, at forty, the oldest man in the crew.

All listened attentively to the weather forecast: a light drift at the moment, virtually nothing, but unstable air flows forecast. Visibility was excellent at twenty-five miles. There was moderate turbulence. The sea was calm. Light showers were forecast for later in the day.

Then the briefing dealt with visits expected at RAF Manston during the day, including two helicopters of the French Air Force which were coming from Rouen and would refuel at Manston.

After that, the details of the royal visits going on around the country for that day were given. Wessex Search and Rescue crews are on a fifteen-minute stand-by by day and an hour's stand-by at night; but this is whittled down to only five minutes' readiness if a flight involving a member of the royal family will be in the vicinity. On this particular day, Princess Margaret was going from London to an air exhibition in Norfolk (Coltishall would obviously watch that), and the Duke and Duchess of York were going from Northolt to Zurich, which just *might* involve RAF Manston if anything went wrong.

Today's pilot took over to detail the tasks for the day. We would go over to the Deal coast and drop a drum, representing someone in the water, into the sea. There would be a cage on top of the drum containing a transmitter giving the alarm signal, called a Pye locator beacon. We would then fly on to the the 450-foot cliffs between Dover and Folkestone, where the winchman would be winched down to a simulated casualty in the cliff face. This would give practice in hovering above the sea, close to a cliff side. After that, we would call up a ferry leaving Dover, and do a dummy approach to the back of the ship. We would then go back and find the drum representing the person in the sea.

It all sounded straightforward and rather mundane, at that stage. It was only when the winchman was explaining the safety drill for real, as I was climbing into my own flying kit, that the dangers in what we were going to attempt began to become *real*. During the flight I would wear, over my flying overalls, something called a monkey harness. This was a broad webbing strap around the waist, attached to a length of wadding like a monkey's tail which, in turn, would be snapped on to a strong rail at the top of the helicopter's cabin. By this means I would be able to walk about the Wessex, look out of the open sliding door and even fall out of it in a sudden squall, without saying goodbye for ever; I would merely dangle beneath the fuselage at the end of the webbing strap.

Suddenly it seemed the reverse of mundane. I was also given a waistcoat-type life jacket secured by three robust buttons to put round my neck and do up. To the left, just beside my rib cage, was a handle made of beads. Pulling this across my body with my *right* hand would inflate a life-saving raft round my neck. In no circumstances should the handle be pulled before I was out of the aircraft and into the sea; otherwise I might trap myself in the aircraft and never be able to get out. Also, at my side, was a pouch containing a distress beacon similar to the one we would be dumping into the sea on the drum.

If I found myself alone in the sea, with the pilot, navigator and winchman all dead, I should, the winchman told me, pull a toggle on the top of this pouch, which would cause a little aerial about a foot long to pop up. Immediately and automatically this would start to relay the distress signal giving my position.

Understood? Understood.

'The conditions today are good,' said the pilot reassuringly. He spoiled it by adding: 'Conditions can change in a moment. It can be very elemental flying. You are well aware of your place in the order of things if you have a hoinking great wind causing problems when you are up against a cliff face. No two vessels at sea are the same, no two cliffs are the same, no two winds are the same.'

The take-off was smooth enough. The Wessex was twenty-two years old but, I was told, probably every spare part on it had been replaced in that time. I was reassured. The Wessex had two engines and flew both, with the remaining engine coming automatically on to full power if the other engine cut out. The maximum speed we would be doing would be 120 knots, about 140 miles an hour.

The helicopter zoomed along the coastline, attracting waves from holiday-makers, at about one hundred feet, until the pilot went into a hover and the winchman fired a Very pistol with an orange smoke cartridge into the sea. He then lowered the red drum into the sea on the end of a grapple hook, disconnected the hook and half-closed the open sliding door – to my relief, as my notebook was being almost torn from my hand in the blast of chilly air through the open side door by which I was sitting.

For the next exercise, the rescue from the cliff, the navigator left his seat up front, beside the pilot, to come downstairs and act as winch operator. He fired the pistol again, sending a flare into the uneven scrub of the high, slightly sloping cliff. The pilot moved the helicopter away from the cliff, circled and came back again as if approaching the 'victim' for the first time. He manoeuvred the Wessex gradually towards the marked spot on the cliff face, constantly prompted by the navigator in the jargon of the trade, in which a degree of movement, called 'one', is about six feet: 'Back one; right, left one; up one . . .'

Eventually, with the helicopter stable and as close as possible to the cliff face, the pilot asked the winchman over the intercom: 'Are you happy to go out there, Bob? If you get into trouble with the cable, we will get you out over the water and recover you over the sea. Happy?'

'I am quite happy to go down there.'

'Fine.'

The prospect did not strike one nervous civilian as exactly *fine*. Pilots will tell you that no helicopter wants to stand still, even in good weather; there is a constant fight to make it stay in one place. The pilot relies on the evaluations of distance by the winchman or, when the winchman is being lowered down the cable, by the navigator. The pilot cannot see where the tail of the helicopter is. He cannot see above him or below him, though the Wessex is fitted with two front headlights and four downwards-pointing lights to improve the general illumination at the scene of a night incident. He relies on the winch operator or the navigator to give him corrections in order to maintain his position, and the rest of the crew rely on him to react swiftly enough to keep the aircraft steady, not allowing a change in wind to smash it against the cliff face.

The prompting of the pilot was constant: 'Back one right; beware, cable is being pulled backwards. Back one. Back one – head and tail well clear. Back one right. Back one. Back one – head and tail well clear. Steady – winchman's almost completed his work. Back and left – steady, steady. Tips and tail are clear. Forward one and left – continue.

Dead ahead one. Forward one and right. Tips and tail well clear, forward one. Well clear on left, above and behind.'

By this time, the winchman had been pulled up, back into the helicopter, on the end of the steel cable.

'Very nice, thank you, Mike,' said the navigator.

'Thank *you*, Peter.'

If they sounded pleased with the job, the pleasure was nothing to my own once the helicopter finally edged away from the cliff face. Hovering is a relative term; there is no such thing as an absolutely still aircraft, except when on the ground; keeping it *virtually* still in the air is a constant battle in good conditions and a sore trial in bad ones, all within a few feet of potential disaster.

The next phase, putting the winchman on the deck of the ferry, *Viking II*, was possibly more difficult as practice than it would have been for real, if the winchman had, perhaps, to bring off a desperately ill man or woman. In a real emergency, the pilot would have got the captain of the ferry to put about and take the course most favourable for the winching operation. But the RAF would not make itself popular with civilian captains or holiday-makers if it caused ferries to put about just for its practice, so our landing would have to be made with the ship sailing in its present direction.

'Hovering' over a ship doing perhaps twenty-five knots is not so much hovering as flying rigidly to the speed of the ship, hoping that its up-and-down movement will not leave the dangling winchman suspended in mid-air one moment and land him on to the deck with a crippling thud the next. Stewards cleared the aft deck of the *Viking II* of the remaining chairs and passengers as the helicopter came up aft of the ship, slowly descending towards the green square that was the deck. Passengers craning to see what was happening were waved back by the stewards. It *seemed* to take quite a long time to get to a position where the winchman could begin his descent at the end of the cable – though in fact it was no more than a couple of minutes before he was back in the helicopter, which veered off sharply to port, well clear of the ferry's funnel. In rough weather, a ship's mast or funnel, rolling from side to side in a swell, can be a nightmare to the pilot and navigator. There is a special drill for running a winchman on to such a ship. It entails initially linking the helicopter to the ship by way of a rope, prior to lowering the winchman. The winchman can then be guided on to the deck by the crew. This ensures that the helicopter spends the minimum of time over the deck – the time of maximum risk.

The crew seemed to feel rather cheated that they were not able to show me this hazardous routine. I confess I was able to hide my own disappointment.

Recovering the red drum from the sea was relatively simply done with the aid of the locator beacon, which gave a constant fix on the position and uttered a wail like a stricken seagull. Once the drum had been sighted, the winchman took over the pilot-prompting again, giving him an idea of how much further forward to go by using the multiples of six feet used in a finer way against the cliff face. 'Direction forward, twenty, fifteen, ten, eight, six, five, four, three, two and right, one and right, steady – winching in.'

The cable with the grappling hook at the end was now lowered on to the drum, picking it up by the cage on its top. 'Winching in,' reported the winchman. 'Twenty feet of cable to come. Ten feet of cable to come. Five feet of cable to come. At the door. Clearing the hooks. Retrieved.'

Having practised three vital techniques – recovery from cliff or mountainside, recovery from the sea and recovery from a ship – the Wessex flew back along the coast at full speed, again attracting waves from people below.

It was only when we were landing that, curiously, the door was completely closed for the first time; and the winchman and I had to abandon our monkey harness and be strapped into individual seats as if we were on an ordinary civilian aeroplane.

I asked the winchman why. 'Because it is at this stage that the helicopter is at its most unstable, as the rotor blades slow down and vibrate. It is at this stage at which the helicopter is likely to keel over.' I could well see that the irony of having recovered, say, a heart patient from a ferry, and then losing him because the helicopter toppled over back at base, sending a piece of the rotors through an open door, would be too much for these dedicated men to endure. But in this case, as is usual, the rotor blades stopped without incident.

'Thank you very much, gentlemen,' said the pilot over the intercom. Within two minutes he was calling them by their first names as, back in the flight building, they analysed that day's sortie with me.

I said that hugging the side of the cliff, with little space between the rotor arms and the chalk of the cliff face itself, must take some concentration.

'We are not allowed to do more than thirty minutes of cliff work on any one sortie,' confirmed the pilot. 'Unless, of course, it is for real, when we have to use our judgment. It is when the weather changes around you that you get into real difficulties. If it starts knocking the aeroplane about, there is nothing much you can do. Once you have committed the winchman to the wire, you have to stop and see it out. Once, we got a report that a girl was injured at the Isle of Sheppey. It was said she had both legs broken, and was in the mud under the cliff. Once we had started to winch her up, we were hit by a squall. As the result of that, flying on instruments would have been dangerous, because of the down-draughting air. The ridge of the cliff was invisible in low cloud, it was night-time, and the lights of the town were being obscured by rain. We had to sit there hovering until I could see just enough over the ridge to go off the hover and get away. You are often left less of a safety margin than you expected and planned for.'

On this particular occasion, the victim's legs were not in fact broken; and, in retrospect, the Wessex crew were not convinced she could not have been taken off the mud by other means. But, at the time, it was a case of theirs not to reason why. A life was seemingly in danger, and they willingly reacted.

The navigator said it was dangerous to think, at the outset of a rescue, that things were going to be hard or going to be easy; you never knew. He thought everyone had in fact done things the validity of which they would afterwards doubt; everyone would have put themselves in positions which they perhaps should not have done, in order to save a life.

It says much for the cool judgment of the rescue crews that very rarely are they themselves injured. The navigator said that, during his six years with the service, he had injured only two winchmen. One had had a finger broken when it got jammed by the winching wire; the other's knee had been injured when he was winched up while he had his ankle wedged between rocks.

The winchman on the sortie I had joined remembered his worst job as a sailing barge which had lost its rudder in the English Channel on a Whitsun public holiday. By the time the helicopter had got there, the ship was rolling broadside on to the waves, with the boom of the sail swinging dangerously from side to side across the decks. On board were four children, four women and two men.

The only way for the winchman to get aboard was to go down a guide-rope 150 feet clear of the swaying mast and boom. He got aboard, and got all the women and children off the boat, leaving the two men with it – knowing that a tug was on its way to tow them into harbour.

Sometimes the winchman's tasks can border on the comic. Locator beacons should become activated only in an emergency, but sometimes they become active by accident, sending out their distress signal giving their exact position. 'I well remember three searches for beacons,' said the winchman. 'One was aboard a yacht in Ipswich harbour and had gone off accidentally; one was in an aircraft in a hangar at Southend airport; and the third was in a factory at Greenhythe, where someone had kicked it over and set it off. I had to run through the streets of Greenhythe in my yellow immersion suit and helmet, trying to find this beacon. What people thought I dare not think.'

Such comic interludes are rare. The seriousness and skill of the RAF Search and Rescue Service are respected throughout the world. The Squadron Leader of 22 Squadron, a soft-spoken Scotsman not given to overt emotion or hyperbole, spoke with assurance on this point. 'We are very much ahead in the search and rescue business. There are good services in other countries – the Swiss and the French are very good. But, generally, in international competitions, if we don't win, then the Royal Navy does. The Swiss and French tend to be heavily specialized. The Swiss are heavily into mountain rescue, and so are the French, because of the Alps. They are not nearly as good as we are over the water. The Germans and Belgians are quite good, but seem to have picked up quite a lot from watching our operations. The Germans have bought Sea King helicopters and have adopted many of our procedures.'

When I left RAF Manston, the day crew still had several hours of their spell of duty to complete. Their conversation was as animated as ever – a tribute, perhaps, not only to their own sturdy temperaments but to the human spirit itself, which sometimes reconciles itself to the necessities of fighting but is perhaps happiest when it can do others tangible good, not harm.

The RAF Search and Rescue Service, leanly supervised by the two co-ordination centres, at Pitreavie Castle in Scotland and at Plymouth, functions twenty-four hours a day, 365 days of the year. All it needs to function, by way of visibility, are clouds no lower than one hundred feet, and four hundred metres of visibility in the daytime. The only thing that will stop it flying, apart from that, is freezing rain, which would probably place the helicopter crews among those needing rescue themselves.

The line between action to save lives and self-preservation is a fine one. Although the men of Search and Rescue modestly and cautiously deny they ever cross it, there is a twinkle behind their denial that says a lot for their dedication to a cause they obviously and unreservedly believe in.

There is no less dedication, though possibly less macho physicality, in that part of the air defences that directly employs semi-civilians in a *fighting* role.

17

Civilians-Plus

> 'I feel my unit, if it had to establish a maritime headquarters in the field because everything else had gone up with an incandescent glow, would be the most capable of doing it.'
>
> – Commanding Officer, 1 Maritime Headquarters Unit, Royal Auxiliary Air Force, Northwood, Middlesex (a Wing Commander and a Chartered Secretary).

> 'They will get rather miffed if you refer to them as *civilians*. They are Auxiliaries. They put in two evenings a week, plus one weekend a month and they have to complete fifteen days' annual continuous training. They range from Securicor people to solicitors, British Telecom personnel, engineers, air traffic control people in civilian life – I know that at one Auxiliary unit there were two engine drivers.'
>
> – Flight Lieutenant, Women's Royal Air Force, at 1 Maritime Headquarters Unit, Royal Auxiliary Air Force.

Early one winter's morning when the cold mists had not yet cleared, a country house with conical towers, on the Hertfordshire–Middlesex border, sprang into animated life amidst the surrounding stillness.

The ICI personnel officer arrived at Valency House, once an Edwardian love-nest but now given over to quite different purposes, closely followed by the secretary of a well-known college. The lecturer at a teacher training college almost jammed shoulders in the doorway with the PhD in geriatrics.

It was only when the chartered secretary of a scientific organization arrived that the day went into top gear. This was because he, at fifty-three and (unusually) with twenty-one years' Royal Air Force experience behind him, was the Commanding Officer of the Royal Auxiliary Air Force Number 1 (County of Hertford) Maritime Headquarters Unit, a formation of 18 Group (based on the old Coastal Command), itself based at HMS *Warrior* at Northwood, a NATO headquarters dealing with various aspects of defence of the Eastern Atlantic and the Channel. ('Don't use the term *English* Channel,' I was warned. 'That might sound too parochial, and you mustn't get the French going bananas.')

The Royal Auxiliary Air Force and the Royal Air Force Reserves are a complex administrative structure which seems to be in constant flux. Sufficient to say that the official briefing I received on it before my visit to Number 1 Maritime Unit (one of three: the others are at Edinburgh and Plymouth) caused the diligent functionary giving the brief to consult his handbook three times and then hesitate over some answers; and that when three officers in a Royal Auxiliary Air Force mess tried to clear up the confusion for me, each contradicted the others. The RAF's borderline with civilian life seems to be a wavy one, even if it works.

What was certain on that winter's morning was that 1 Maritime Headquarters Unit was anxious to point out that it was *not* the equivalent of the Territorial Army. It was made clear that it was pumping technical expertise from civilian sources and civilian individuals into a very sophisticated game of electronics and radar that watched the Warsaw Pact naval forces with the vigilance and precision of professionals – and perhaps bearing even greater strains because they had to integrate their hi-tech work for the Royal Auxiliary Air Force with their civilian jobs and lives.

1 Maritime Headquarters Unit does its most secret work in the NATO complex just up the road, where visitors are

screened very carefully indeed; but it is administered in the rather more relaxed atmosphere of Valency House. While I was waiting there for the arrival of the Commanding Officer, I leafed through two enormous leather-bound books of cuttings brought for me by the Adjutant, a brisk young lady with formidable spectacles and the rank of Flight Lieutenant in the regular Women's Royal Air Force. She was one of the seven regular RAF people who form the permanent seven-days-a-week administrative backbone of the unit, seeing the Auxiliary civilians at weekends and on two evenings during the week.

In the cuttings book were copious pictures of officers with Royalty and local bigwigs, receiving *this* and handing out *that* – which, though pleasant enough, might or might not have direct bearing on the defence of the realm. There are those anti-military civilians who regard the auxiliary forces as a way of attaining weekend self-importance by playing at military games and military ranks; and I began to wonder whether these records of a bulging past would give such people ammunition. My fears were not eased by some of the menus pasted in the books – those of ceremonial dinners and guests nights, also attended by Royalty and local bigwigs. I would have been delighted, on 21 November 1987, in honour of a distinguished Air Commodore, to drink Oppenheimer Krötenbrunnen, Côtes du Rhône, Graham's Port and Duke of Clarence Fine Old Malmsey to wash down the smoked salmon and venison; but would it have strengthened NATO more than it weakened the Exchequer?

It *didn't* weaken the Exchequer, I was briskly told. On such social occasions, the Auxiliaries paid for the food and drink *themselves* – at around £20 a head, which was cheaper than it would otherwise be for them only because they could use RAF venues at no extra cost to the service.

This seemed a significant arrangement. Of the many Auxiliaries I talked to, most emphasized that it was the thought of *giving* something to the country that was the attraction, rather than receiving the pay, which in many cases would not begin to compensate for the domestic sacrifices and civilian job compromises entailed.

'Your company chairman might be sympathetic to your being in the service, because he may be old enough to remember the last war,' said one officer at Northwood. 'But your immediate colleagues might not be so sympathetic. They might say you were doing it just for the cash benefit. You would be surprised at how narrow-minded and parochial people can be if they want to be.'

In view of such pressures, it may be surprising the Royal Auxiliary Air Force holds together as well as it does. It was formed in 1924. There are support flights in the three Maritime Headquarters units. There are the six Air Force Auxiliary Regiment field squadrons, at Lossiemouth (Scotland), Waddington (Lincolnshire), Honington (Suffolk), Marham (Norfolk), Brize Norton (Oxfordshire) and St Mawgan (Cornwall). There is a movements squadron at Brize Norton assisting in the deployment of aircraft personnel to Europe, an aero-medical evacuation squadron at Hullavington (Wiltshire) and a squadron based at Waddington. There are four Royal Auxiliary Air Force flights making up a Home Defence Support Force.

A clear structure of command, I was told at the time of my visit to Northwood, was emerging; but at present there were separate units almost running themselves.

There could be psychological strengths as well as administrative weaknesses in this fact. 'At the weekend I tend to get 80 per cent of my strength coming in,' said the Commanding Officer, a man with the youthful face of a serious and clever schoolboy, who said it was his immediate ambition to buy a Morgan sports car and drive along the course of Julius Caesar's military campaigns. This percentage of attendance certainly compares well with that of the Territorial Army.

The Commanding Officer was himself an example of the sort of enthusiasm that holds the Auxiliaries together. He had been in communications and Intelligence duties in the RAF for twenty-one years until he left in 1973. He had a complete break for eight years; and then he became secretary to a charity appeal, where he met the then head of the Royal Auxiliary Air Force. It reactivated his interest in the RAF. He joined the Auxiliaries.

'When I had been in uniform again three months, it began clicking back into place again,' he told me. 'I have got this command for four years, of which there are another two years to go. You can stay in the Auxiliaries until you are sixty, but whether there will be another job for me, I don't know. The one thing I have found, in command, is that it is another job in itself. If you have got a full-time job in civilian life – and that applies to the true Auxiliary – you have got all the strains and stresses of holding a major executive post as well as the job you are doing in civilian life. You can be in your office in London, but you have got to be ready to deal with things here at the drop of a hat if necessary because you are the Commanding Officer, and only you can advise.'

I was not surprised to hear that this officer, among others, tried to keep his two roles as rigidly separate as possible, asking telephone callers to his London office always to call him 'Mister' rather than 'Wing Commander' and to keep their calls as short as possible.

The number of training days that Auxiliaries are expected to serve during the course of a year is fifty-eight. Not sufficient from a command point of view, said the Commanding Officer. The regular RAF was not available at the weekend, when most Auxiliary action took place; this complicated the business of getting stores, and so on.

'There are times when I just have to take a day off and get my butt down here,' confessed the Commanding Officer. 'I have to balance everything I do between what the RAF can make available and the requirements of 18 Group. If I require extra equipment or funding I may have to ask the Territorial Army Volunteer Reserve for help.'

Did he, like several Commanding Officers of regular RAF stations, have a fixed budget of his own, which he could apportion? 'This is being discussed at the moment. If it comes about, and there is an experiment, this unit will probably get that experiment because I have personal experience of handling budgets.'

As things stand, the right state of mind about NATO in general and the Americans in particular is possibly more important than the manipulation of finance. It certainly would be if the unit had to assume its war role.

'We work closely with the United States Navy Reserve,' explained the Commanding Officer. 'We must be able to mix with them and we must have the right kind of personnel to do that. You don't want anyone who doesn't like Americans, for instance. You don't want anyone who is a bit of a griper – a grumbler. If he has his pet likes and dislikes and brings them into an area where you have to have total integration and co-operation (otherwise you don't get the right result) you could have a disaster on your hands. The chap might produce preconceived results. If he is an Intelligence officer, he might say, "I don't like that report from the Americans, it must be useless", just because he didn't like the Americans. Similarly, because the French are involved in some big exercises, you must have people with good French who like the French. You can't have people who might get a signal and not understand it or who, not understanding it, would simply think it of no importance and put it on a shelf.'

Training of Auxiliaries therefore tends to include service on as many overseas stations as possible, as well as on their own units. In war, Northwood would send personnel to reinforce Gibraltar, where there is a huge communications centre built inside the great rock itself, safe from any high-explosive attack. In peacetime, communications personnel go regularly to Gibraltar for exercises. Occasionally they go to Portugal. Personnel visit the other two Maritime units and also attend training courses at the British Army Intelligence centre at Ashford in Kent.

If, in these functions, the Auxiliaries have to play a partly diplomatic role, they have to play the same role with the local communities around their home bases. 1 Maritime

Headquarters Unit has been adopted by the Worshipful Company of Butchers.

This might, at first sight, seem a terminologically unfortunate title for an organization associated with a fighting force. But there is a logic to it. The unit is the only Auxiliary unit in the Greater London area; and the Worshipful Company of Butchers adopted 600 Squadron in the immediate post-war years. Today the Commanding Officer is often expected to dine with the Worshipful Company, and its leaders are frequently invited out to Northwood.

Diplomatic niceties must always take second place to solving technological problems. Front-line units like Gibraltar, said the Commanding Officer of 1 Maritime Headquarters Unit, were always trying to find ways of improving their techniques (which would help to bottle up the Warsaw Pact fleets in the Mediterranean in time of war); but they were up against cost factors.

'What is absolutely necessary,' said the Commanding Officer, 'is that front-line units like Gibraltar should identify the areas where they require reinforcement. They have to be quite precise in this identification. It is no good asking me to supply four motor transport drivers and when I send them four, they turn round and say, "Oh, we wanted drivers trained on such-and-such a vehicle, and these aren't." That is the result of the receiving people not precisely identifying their requirement.'

I asked if this whole sensitive area was really suitable for the involvement of civilians such as the Auxiliaries, bearing in mind that I had been told: 'The task is to keep an eye on Boris when he is coming over the North Pole with his submarines and capital ships, to locate them and, in war, to destroy them.'

The answer was that regular services would need to be augmented in war, and that Auxiliaries were the *only* answer as well as the best one. There would be an increasing pressure to recruit Auxiliaries in future as the population got older and older, with its youngsters further and further away from those memories of war that often persuaded men and women to join the part-time military. People like the woman Squadron Leader at Northwood, who was shortly retiring after forty years to devote herself full-time to cat breeding, were becoming fewer and fewer, I was told.

'If we put it in rather plain terms,' said the Commanding Officer, 'there are 365 days in a year. The normal person, in his civilian job, works 220 of them. A Royal Auxiliary Air Force person who is reasonably committed uses the remaining 145 days in his capacity as a reservist. But when we are on duty, despite the strains, there is little if any difference between the experienced Auxiliary and the experienced regular. The reason is that we do all our work, or most of it, in the same place over a number of years, whereas our regular counterparts keep moving jobs every two years or so. My problem is not expertise: my problem is bringing new blood in.'

The strength of the Auxiliaries lies chiefly not in the rank of retired RAF people but in people who may not have had *any* service in the RAF, but are prepared to train in its daunting technologies.

I met quite a few at Northwood. One was a forty-seven-year-old Flight Lieutenant, the already referred to secretary of a well-known college. He joined the Auxiliaries as a second aircraftman – 'a rank so low it now doesn't exist' – after the RAF itself rejected him as aircrew but told him he would still have to do his National Service. He joined the Auxiliaries to help guarantee that he got into the RAF in some capacity when he was called up for National Service.

'In the end I never did National Service because it was gradually being phased out. I was never called,' the tall, smooth and academic-looking officer told me. 'Although when young I thought there was no point in being in the RAF unless you did the actual flying, I see life in the Auxiliaries as better in some ways than the life of the RAF now.'

At the time we spoke, the officer was a Royal Auxiliary Air Force Regiment officer, as he had been for nine years. He was handling operational defence training, and also helping the regular adjutant to do administration.

Did he enjoy working under a woman? – 'Not *under* – *with*!'

From unit to unit, it is never predictable whether the Auxiliary will find himself working with many women, few or none. The proportion of expert regulars to expert Auxiliaries is also unpredictable. On some units, 30 per cent or more of the staff will be regulars, whereas at Northwood a small staff of regulars are responsible for administration while the large force of Auxiliary experts are responsible for the technologial know-how.

The Squadron Leader at Northwood responsible for Operations, with four flights underneath him, to be trained and held in readiness for their roles in wartime, had a great deal of military experience to draw on. A short, stocky former pilot in the RAF with twenty-five years' flying to his credit, he did his primary training on Shackletons and then went to the aircraft firm Hawker Siddeley for the development of Nimrod. From there he went to the Staff College for career high-flyers and thence on a two-year exchange with the Royal Australian Air Force as an instructor in its anti-submarine school. Five years in Strike Command, and working on NATO plans in Gibraltar, saw him nearing the end of his RAF career; he had moved around all his life, and his wife wanted a more settled atmosphere.

It did not mean a severance from aircraft. He took a job as an operations officer with the Civil Aviation Authority, involved in flying-crew licensing.

Did he ever feel his life had been *too* full of aeroplanes? 'I can't get *enough* aviation! I still fly with the Civil Aviation Authority; I am a civilian flying instructor. My problem is that there are not sufficient hours in the day. Clearly I am next in line to take over command of this unit, but I don't believe that my civilian job would allow me the time that would require. It needs two days a week, full time, and my job won't let me have that time off. I have considered leaving my full-time employment and becoming Commanding Officer here and doing a part-time job as a flying instructor, or just staying in this job as Squadron Leader Operations until retirement. I have to make a decision; and I think it will be driven by pension considerations in the end. I will stick to my civilian job unless a miracle comes off.'

Like winning the football pools? 'I don't think there is such a miracle.'

The Squadron Leader Operations said a trifle defensively that it was always made clear when interviewing would-be recruits that their primary responsibility must be to their families and livelihoods. If, after they joined, there was some problem in meeting the requirements of the service, they were asked what the problem was. Usually it could be solved – such as by giving them six months' leave to sort out family or business problems. After this, if the difficulties went on, 'We have to invite them to leave.' There was a continuous turnover of people, especially in the junior ranks. Once they reached Corporal, it was pretty clear they were in the Auxiliaries to stay.

Some Auxiliaries with purely civilian backgrounds are not those you would expect to find in *any* military context. A thirty-seven-year-old teacher, who was a Sergeant engaged in analysing information on Warsaw Pact naval forces for an Intelligence flight, wore the sort of fair walrus moustache that may have been a military hallmark in Kipling's time but has more often, in our own, been the hallmark of student and academic anti-Establishmentism.

How had he come to be connected with the Auxiliaries? At college, he told me, he had studied languages, and his tutor in geography had been an ex-RAF man who worked a lot about photographic interpretation into the geography course because he happened to be personally interested in the subject. As a student, the Sergeant had also made money on vacation by working as a security guard.

All three skills were of obvious value to him in his present Auxiliary role. But he told me he had first approached the Territorial Army. They expressed an interest in him, but he had also approached the Royal Auxiliary Air Force: 'I think I decided on them because it was slightly more civilized and urbane, a little less rough-and-tumble. The idea of very hard conditions for two weeks, and climbing around, is all right for a time; but in the long term it is not necessarily the most interesting thing. The Royal Auxiliary Air Force is slightly more cerebral.'

When he first joined the Royal Auxiliary Air Force, eight years previously, there was still a perhaps dominant element of anti-military feeling in the common rooms of colleges – though things had certainly changed since. I asked him what his friends in education had thought of his move.

'I think they were interested. They thought it was something unusual. There were certainly a number of misconceptions to clear up. People tended to think of it very much in terms of being similar to the Territorial Army, because the best way one can explain it is to say it is the Air Force equivalent of the Territorial Army. You then have to explain it is not lying in ditches all night. It is odd that though there is an element of personal defence training, doing camouflage and firing rifles, there was not, in fact, this training when I *first* joined, though everyone tended to assume that that was what it was all about.'

But judgment, he insisted, was really the name of the game. Sometimes, when analysing information about the Russian fleet, you thought, 'Hello, people haven't noticed *this*' – and you took care to point it out. Any deviation from the expected had to be thought about and analysed.

I wondered whether this academic man would be more comfortable in the officers' mess than in the sergeants' mess; 'I have already applied for a commission.'

Did he expect to find that more congenial than the people he was now mixing with? 'Answering that question takes one's ability to be diplomatic. It is congenial in that one can now work with non-commissioned people and not feel in any way superior; but, to be realistic, some of them would not be people one would choose as friends.'

The sexual barriers may be rather less than the educational ones. At Northwood there are a number of women Royal Auxiliary Air Force officers engaged on Intelligence duties, which are demanding tasks, especially for very young women. One twenty-two-year-old student of a polytechnic in the Midlands had been in the Auxiliaries for eighteen months and had been commissioned as an Acting Pilot Officer (W) only two weeks previously.

'I haven't really found it difficult, the fact that I cannot really talk about my job,' she told me. 'I work on the Intelligence side of 18 Group alongside the regular RAF and the Royal Navy. I can talk to other people in the unit about it. With anybody else, I have to keep it to myself. But I have other interests, and my outside friends are going to be more interested in them anyway.'

She shared a house with three other girls in the Midlands, she told me, and hoped to join the RAF in the administrative and secretarial branch for a short service commission of five or six years. When she came out of the regular RAF, she would like to come back to the Auxiliaries. Even if she had married by then? 'That would depend on the circumstances at the time.'

I spoke to two women Corporals at Northwood, one of whom left me with a flea in my ear for daring to ask her whether she could combine the Auxiliaries with marriage. 'This is a rather personal line of questioning,' she retorted. 'I don't have any plans in the foreseeable future for getting married.' Twenty-nine and a solicitor, she had been a member of the Territorial Army at university, but found that, when she took a job, she had to work on Wednesday evenings, which had been her training night. In the Auxiliaries, she said, she had been trained in communications to be a telecommunications operator. She knew about computers and radio. Her only complaint (apart from being asked 'personal' questions by civilians) was that sometimes she felt that the Maritime Headquarters unit was not using these skills to the full and was imposing too many 'mundane tasks' on trained people like herself.

But a fellow woman Corporal in the same line of work said that the job was 'quite civilized and great fun, and very good from the leadership point of view'. She said that applied in her civilian job with a building society as well as in the Auxiliaries.

'I am a telecommunications controller,' she said. 'It means supervising message-handling of any sort – radio, Morse, telecommunications or computer. You supervise people watching screens and deal with any rejects they cannot handle – a reject is when a message isn't in the correct format. It definitely helps me in my civilian job. They think quite highly of the Auxiliaries. We are trained in personnel management, rotas, interviewing and discipline – and all those can be carried over, to some degree, into a civilian job.'

This particular Corporal said she felt the Auxiliaries used women equally with men – 'If there is a guard duty, you do it the same as the men.' What they would do in war was always a consideration, and she thought about it from time to time. She thought of it chiefly in terms of movement: people on the base would be moved elsewhere, and the Auxiliaries, women as well as men, would move in to take their place. The only time it happened that she could remember was for the Falklands campaign; but it was a possibility that had always to be borne in mind.

There are not many black or brown faces in the Royal Auxiliary Air Force, but there are some. At Northwood I found one twenty-one-year-old aircraftman who was born in Britain of Mauritian parents. He had a military background: his father had served twenty-two years with the Royal Pioneer Corps, and his older brother had done twenty-two years in the Royal Navy. He himself left school at sixteen with six CSEs, did a craftsmanship course in electronics with a large electrical firm and then moved to work in the drawing office of an electrical components company.

He said he had been a cadet but had left the Air Training Corps when he was in training for his civilian job. But when these pressures on him became less, he realized he had too much time on his hands. He went down to his old squadron one evening, and his former Warrant Officer told him about the 1 Maritime Headquarters at Northwood.

His civilian employers, he said, were quite impressed with his increased versatility since he had joined the Auxiliaries. 'I am becoming more of a tradesman in the technical field. They are quite impressed I have an outside interest and am getting involved in the social activities and events. You meet people from all walks of life and you learn to get on with people better. All that is good for your civilian employment.'

I asked the Mauritian aircraftman whether he had ever encountered colour prejudice. Not among his colleagues in the Auxiliaries, he replied; but occasionally people would jeer at him in the street that someone like him was not fit to wear the Queen's uniform. Some West Indians and Indians, he said, regarded a uniform as threatening. 'Some West Indians don't understand how people can give up their time to serve Queen and country, and possibly some Asian minorities and white people are the same.'

Certainly his feelings could not be put down to personal touchiness. Though a trained technician, he arrived for our conversation wearing dungarees covered in white paint. He had been helping his colleagues to repaint the shed in which they trained. I asked him what he did to relax after his civilian job and his Auxiliary training: 'Oh, I have other activities – like aircraft modelling.'

He was obviously a deeply dedicated soul – and one with a dark skin who said that his favourite newspaper was the *Daily Mail*, and that he was non-political – though he had voted Conservative because he liked some of the things they had achieved, though not others, like unemployment.

What you find in the Royal Auxiliary Air Force (and 1 Maritime Headquarters Unit was just a taste of an organization which also includes Royal Auxiliary Air Force Regiment field squadrons, movements and aero-medical evacuation squadrons, the Light Anti-Aircraft Squadron and the support or defence flights) is unpredictable; certainly not always in accord with class, sex or race stereotypes. Plainly its unpredictability is one of the attractions to those who turn out weekend after weekend at places like Northwood, often working underground in secure but sombre environments.

As a rational taxpayer one is, of course, pleased to know that the whole of the Air Cadets, the Auxiliaries and the Reserves are being rationalized; but there may be losses of that quirky human unpredictability that probably *appeals* to civilians rather than repels them. At any rate, when the Commanding Officer of 1 Maritime Headquarters Unit advertises in the newspapers an invitation to apply for commissions in the Royal Auxiliary Air Force, he usually gets about eighty enquiries, interviews about twenty of the enquirers and takes on the half-dozen he needs.

Organizational oddities evidently do not put people off when the urge to do something extra for Queen and country – elsewhere often derided – is upon them. It is good for the image of the Royal Air Force in the civilian community – as is another, highly melodious, Royal Air Force activity.

18

On Public Show

'It may be more difficult if you are a Jaguar pilot – or in any one of the more aggressive roles – to generate good public relations with the public, because there may be an antagonistic look about the service in these roles. Whereas in the Music Services you can promote the Royal Air Force with the sounds of the Broadway era, Gilbert and Sullivan or Glenn Miller – it's gorgeous.'

– Principal Music Director, Music Services, RAF Uxbridge (a Wing Commander).

'Do *not* PLEASE, talk about bandsmen in the RAF! The British *Army* has bandsmen. The Royal Air Force has *musicians*.'

– Officer in Music Services, RAF Uxbridge.

The goat made a lasting impression on the Royal Air Force Bandmaster, not to mention the polished parquet floor. The Bandmaster was about to raise his baton for a ceremonial fanfare in the town hall of a German town near Brüggen, where smooth RAF–German relations are vitally important to the RAF base there. Instead, he found himself suddenly, spontaneously and solemnly presented, by a local official, with Mr Bruggen (the goat) as the RAF Brüggen mascot.

Unfortunately, there was no book of instructions with the goat, nor was there a laid-down protocol for such a situation. So the flustered Bandmaster, falling back on the British stiff upper lip, tried to hold the goat's rope in one hand while striking up Wagner with the other.

The goat had other ideas. It proceeded to jump up and down, to slip and slither on the shiny floor and otherwise to misbehave itself while Wagner was conducted in a way not heard or seen before or since, and the audience vigorously applauded the extra entertainment. To the Bandmaster it seemed hours before a merciful subordinate managed to drag the protesting goat away from the podium.

The Germans loved the robust joke rather more than the Bandmaster, who told me afterwards: 'At one stage we thought the only possible way to take the goat back to base was to put it in a bass drum case.'

Mr Bruggen was the successor of a previous goat which had been given to the band. The RAF vowed that this time, instead of trying to keep him on base, Mr Bruggen would be sent to a local zoo – which undertook to house and feed it except when the RAF wanted it for a specific ceremonial occasion of vital importance to Anglo-German relations.

There can still be one or two moments of slight tension at such occasions. A German mayor may suddenly, in a speech, recall being a prisoner of war in Britain during the war. Another may recall his first sight of the RAF as a child – seeing it overhead on its way to bomb a German target. At such moments of tension, the presence of a goat tends to remind the company they are at least part of one species and the goat another – one moreover likely to chew up the sheet music without warning, an aberration few humans, apart possibly from Hitler, have allowed themselves.

The Central Band of the Royal Air Force, the Band of the Royal Air Force College, Cranwell, the Band of the Royal Air Force Regiment at Catterick, the Western Band of the Royal Air Force at RAF Locking, Weston-super-Mare, and the Band of the Royal Air Force Germany, at Rheindahlen, are *all* an important feature of the face of the RAF as seen by civilians whose goodwill – as neighbours or as taxpayers – may be vital to the RAF.

All are administered, together with the Royal Air Force School of Music, from RAF Uxbridge, a station conveniently near London which acts as home base and which also handles all the paperwork for RAF personnel who are abroad. The Music Services occupy a modern (1969) black and white brick building which (except for the RAF prints on some walls) could easily be mistaken for broadcasting or recording studios rather than a military establishment, and in which the conversation tends to be more about recording contracts than spit and polish.

The Central Band of the Royal Air Force, with seventy-six men, is the prime asset of the Music Services, playing at ceremonial occasions for all three of the armed forces and for civilian organizations when military requirements permit. About a fifth of the work is done for civilians or service charities, but RAF musicians avidly point out that other nominally military occasions, like passing-out parades, include a great many civilian friends and relatives as spectators.

Trying to keep track of the Central Band of the Royal Air Force is like trying to jump on the back of a kangaroo. Its Director of Music, a Squadron Leader with the sympathetic face of a northern comedian and, like so many service musicians, a background of playing with the Salvation Army as a child, told me he was just off to play for the Fifth Airborne Brigade of the Army's Parachute Regiment. The band would play for a reception and Beat the Retreat, better known in the RAF as the Sunset Ceremony.

There was infinite variety, said the Director of Music, in spite of the fact that the Central Band was a full orchestra but *without* strings. As the members of the orchestra tuned their instruments in the corridors while waiting for one of the five buses on the base, the director pointed out that most of the Tchaikovsky and Beethoven symphonies had been adapted, and that some members of the orchestra had written their own music especially for an orchestra without strings.

Such as himself. He composed a series of Motto Marches, based on the four main operational stations in Germany. He took the motto of each station and, from the general role of the station and the location of each station, tried to inject that sort of feeling into the music. This has been issued as a commercial recording.

RAF Gütersloh's motto, as the most forward operational station in Germany, is Ramparts of the West. 'It suggested walls, and so there had to be something solid about the sound, something that could not be broken. Something with a slightly historic atmosphere, a sort of Rome.'

Brüggen's motto is To Seek and Strike. There had to be a feeling of seeking, at the same time electronic sounds and instruments, and a sound which suggested the eventual finding of the target. The feeling had to be not incompatible with sheer music, but it had to mark the achievement of the machine and the person in finding the target.

A Firm Fortress is the motto of RAF Laarbruch. This is by the Dutch border, and rather pastoral. The music had to sound like part of the countryside and at the same time have something of a different, military tone to it.

Finally, Wildenrath, which carries priority passengers and freight between stations, has the motto Always Ready. 'There had to be a suggestion of alertness, but at the same time great stability. Writing music takes a great deal of time, and a lot of it gets done, though people are not pressured to do it,' said the Director.

For an orchestration of a twelve-minute suite culled from Holst's *The Planets* suite – which lasts three times as long – this Director of Music reckoned he had spent a hundred hours of work. I was glad to hear that the music was registered with the Performing Right Society – which makes royalty payments to composers on a centralized basis – in the name of the officer personally, *not* that of the RAF, since such creativity cannot rationally be considered as under the control of the service. But musicians insist that such six-monthly payments are small. One told me, 'I got a cheque for £9 today.' Another confessed, 'Mine was only for £6.'

The RAF's successful interaction with civilian life can be seen most readily in the Music Services. Musicians will not discuss how much they make from civilian sources, but the fact is that musicians can accept civilian engagements and keep the fees, on the strict understanding, known to the customer as well as the musician, that if a military need shows itself, the civilian engagement will have to be dropped. Average earnings in the year 1987–88 were less than £4 a week, subject to tax.

'We try not to get into the business of cancellation if there is a chance it will bring the RAF into disrepute,' the Principal Director of Music (in overall charge of all bands and the school) told me. 'And in the event of war, you must remember we have a military role, which we exercise twice a year. If indeed the sad time came, we would put our instruments away and, because musicians are fully trained in firing weapons, we would be used as guards.'

Any bright MP who suggested in Parliament that they are somehow ripping off the system by undertaking private work would be sticking his neck out. The RAF believes it is getting and giving value for money in having such a fine body of musicians to act in an ambassadorial role.

The Principal Director of Music, a tall, gentle-voiced Wing Commander from the north of England, who first played in the Salvation Army band when he was seven, pointed out that the ambassadorial role started at the top. He was shortly, he told me, going to Australia to lecture and guest-conduct civilian bands. He was going to have a look at the Chile Air Force band, and perform other foreign engagements.

To my civilian eyes, he was himself a useful illustration of the way the RAF tries to use talented men who might, if too submerged in military spit and polish, prefer a life in civvy street. He auditioned for the Central Band of the Royal Air Force when he was seventeen, having been an enthusiastic listener to its radio performances. He had worked on tugs on the Tyne, and for a local builder. Then a cousin who had completed a Bandmaster's course in the British Army encouraged him to try service life.

There is none of the usual officers/other ranks divide in

the Music Services. *All* candidates go through the school of music as student musicians, unless they are already qualified, and some have the option of later going back to the school on a two-year Bandmaster's course which qualifies them to become conductors and can give them officer rank. There is no pre-arranged 'élite' which goes on a separate course as of divine right. The Principal Director of Music went through this machine and emerged a formidable euphonium player as well as an administrator who masterminded the Squadronaires orchestra back into existence.

The Squadronaires are a legendary name to anyone whose memory goes back to the Second World War, when they were formed to play at dances and other events at which energetic light music could improve morale. They were drawn from the same corps of musicians who formed the Central Band (more military and serious in character) but had a distinctive style of their own. They made numerous gramophone records and did many broadcasts.

Immediately after the war, many people understandably wanted to forget everything about it. The band eventually declined with a change in public taste. This was after its take-over by civilians, who were up against the fact that the Squadronaires were too old to be novel and too new to enter the nostalgia stakes. It disbanded itself, leaving stranded a tradition established by many dance musicians of note who had joined the RAF in the war. They included Ambrose, George Chisholm and Steve Race, and were comparable with the 'serious' musicians such as Norman del Mar, Denis Matthews, Harry Blech and George Malcolm, who had also been with the wartime Squadronaires.

'I took over as Principal Music Director in 1963 and encouraged the big band era,' said the Principal Music Director. 'We have got some very fine dance-band players in the Central Band. There was generated in the Central Band a wish to form a big dance band. Because of the reviving interest, it felt natural that we would wish to reform the Squadronaires after thirty years. I approached the Ministry of Defence and said there was a lot of nostalgia about. The

Ministry had no objections at all, but said it would be a courtesy to ask the surviving members of the Squadronaires if they had any objection to their name being re-used. The response was tremendously enthusiastic. They even came along and helped us launch the re-formed Squadronaires.'

Some members of the RAF thought it a bit unsporting that a civilian group called themselves the New Squadronaires. 'But *we* are the Royal Air Force Squadronaires,' said the Principal Music Director. 'And we prefer the Royal Air Force in the title to be spelled out in full.'

Such little touches of fussiness do not obscure the fact that life in the Music Services of the Royal Air Force blends easily with that of civilians partly because it has a lot of civilian thinking about it. The rank system is on similar lines to that of a civilian orchestra. Each orchestra section has a specialist performer, a position filled by a chief technician. Each section has a principal, who is a Sergeant. The sub-principal is a Corporal. The rank-and-file players are junior technicians.

I asked the Principal Music Director whether military ranks could always reflect accurately the artistic calibre of the player. He took the point at once. 'Ah! We get round that. In the Central Band personnel at the moment, the man who plays principal cornet is a Corporal. His second leader – i.e. the chief technician – recognizes that the young man is exceptional in the solo field, and plays underneath him. It is still the senior leader's responsibility to make sure the section operates professionally in all aspects, but the actual soloist's chair can be filled by the younger and junior man, who will eventually be promoted anyway.'

In parts of the other two forces, it sometimes seems that even in the Christmas amateur dramatic pantomimes, it must always be the *officer* who plays the *hero* and the Sergeant who plays Widow Twankey. The RAF has compromised with nominally more egalitarian times by accepting the fact that competent musicians may well, in their own minds, put music first and service life second. It has come to terms with this fact.

At Uxbridge I met a twenty-five-year-old junior technician – a clarinet player who had been in the RAF just four months. He had just left the school and become operational. Reading between the lines, I thought he had possibly regarded the school life, at age twenty-five, as being a bit much, especially since he had previously attended a music college in civilian life. He admitted that he was glad to be out of the school and actually at work.

Why the RAF? 'I think their bands are of a higher standard than the other military bands and there are more musical duties than there are in the Army or the Navy's Royal Marines.'

Which was dominant in his mind, music or the service life? 'Music is really the dominant thing. But now that I am in the service I enjoy that side of it a lot.'

So far, he had done only four jobs, for one of which he had been paid. This was when the band spent a week giving concerts in St James's Park; most military bands give a week's music to public parks in turn. It gave one concert in the afternoon and one in the evening for each of the seven days, with twelve items per concert, some 170 different items.

Letting the members of the band share in the proceeds of such concerts for civilians seems a wise policy. In many ways, the musical knowledge required by a member of an RAF band is the equal of that required by a member of, say, the London Symphony Orchestra or the Philharmonia; and the grasp of a wide repertoire must be, if anything, more in evidence. The junior technician said he liked all sorts of music – though classical was his main choice, with Elgar, Mozart and Wagner to the fore. A senior officer who had overheard our conversation put in: 'Purists are hypocrites, in my opinion. Anyone who says one composer is all there is to it is very narrow-minded.' This aside could, I felt, stand as the motto of the Music Services.

The RAF insists on ordinary physical fitness in its musical ambassadors but, diplomatically, does not apply fitness rules quite so strictly. It is perfectly understandable that the

Army, which requires young recruits to jump out of dummy windows, doing a forward roll as they land on their backs in a sandpit, should insist that a recruit must be ready for anything: ultimately, an army is still about hand-to-hand combat. It is equally understandable that the RAF, whose function is rather different, should see the need for fitness differently.

Especially in the case of its *musicians*. I gathered this from, among others, a forty-three-year-old Flight Sergeant who had just graduated from the school of music's Bandmaster's course (i.e. for Warrant Officer rank) and was celebrating with colleagues.

He had always wanted to take up music as a career, he said, right from the time when, aged seven, he had blown the tenor horn in a Salvation Army band in his native Liverpool. He had applied to the Army but been turned down on medical grounds: he had one flat foot and he was 'slightly overweight'. The Army told him he wouldn't do for a bandsman, but might be all right for a cook or a medic. Goodbye, thoughts of the Army! For four years he had an office job, which 'bored me silly' while he developed his trombone playing, which he had decided was the way he was going to get his living *somehow*. He applied to the RAF and was accepted, to his surprise and relief.

He was then twenty-one. Why had he gone back to school for a Bandmaster's course – which meant he could assist officer-conductors or take over the conducting himself – after so long in the rank and file? 'I have always aimed to get this far. I think many people do. Their ambition is to get as high as they can get. The Bandmaster's slot is a very satisfactory one. You have reached the top of the tree, although you have not climbed a different branch and gone for a commission.'

I asked him why he, a brisk man socially quite at ease with officers, hadn't chosen to become one. The answer allowed itself a frankness that might not have come so easily in either of the other two armed services, or in other departments of the RAF.

'Because I am not a whole-hearted serviceman,' he said – in the presence, incidentally, of senior officers. 'I am not prepared to give up my devotion to family life and all the rest of it. As a commissioned officer you have got to give up these things – and I am *personally* programmed. It has never attracted me at all. One of the attractions of the music side of the service life has been the non-routine factor, the fact you never work nine till five. I don't object to working long and uncertain hours sometimes, but the responsibilities of being an officer are not for me.'

That attitude could have been symptomatic of the 'happy bandsman' syndrome said to afflict some British civilian symphony orchestras, who are said to lay down their pint pots once in a while to rehearse a few notes. I was assured by another of the Directors of the Music Services (a compact, gnarled and alert little Squadron Leader with over thirty years as a musician) that this was far from so.

As we sat in the officers' mess at Uxbridge, almost wholly used by men in civilian clothes, I drank tomato juice, and he drank pineapple juice. We both contented ourselves with a single cheese-and-ham sandwich. 'Musicians are not drinkers more than any other class of people,' he insisted.

And the complexities of the building of the Music Services, and its equipment, showed that this *needed* to be so. Apart from a rehearsal room large enough for a full orchestra to make a recording, there were several smaller sound-proofed rehearsal rooms no bigger than a domestic box room, and what is nicknamed the Playroom – a large room with a collection of about one thousand gramophone records, plus compact discs, audio and video cassettes, and sound recording and video recording equipment. I started to plough through some of the gramophone record collection, assisted by some of the staff. Soon we began talking about Simon Rattle's respective mettle in Stravinsky and Rachmaninov in a way that would put any civilian completely at his ease and enable him to forget matters like military discipline and missiles – things about which the civilian mind is apt to be ambivalent.

The Central Band of the Royal Air Force and the other RAF bands are certainly a bridge to civilian minds – and not only in the immediately obvious sense of being heard and seen at Wembley international sporting events, Lord Mayors' Shows, Edinburgh military tattoos and tours of Hong Kong and Japan. Music has indeed proved a powerful emotional and intellectual means of promoting unity and goodwill.

This is helped by the fact that the musicians have a distinctive uniform. On the whole, the RAF is content with its image of a no-nonsense service that is not hidebound by manic regimental ritual, mess practices and exotic uniforms. But sometimes some officers and airmen lament the fact that the service does not have a resplendent full dress uniform – at least for ceremonial occasions.

They point out that, on the whole, when the RAF uniform has been modified, it has been modified in favour of greater simplicity, not that greater flamboyance which might catch the public fancy and perhaps persuade more young people of both sexes to join.

One officer at an operational station, showing me the station silver and trophies, of which there was no lack, stopped by a glass case. It contained a dummy wearing the RAF uniform of the first Station Commander, who had gone on to be an Air Commodore at whose death the widow had sent his dress uniform to the station for display. There was a high collar to the jacket decorated with real gold wire; there were seven highly polished brass buttons on the front of the jacket and two at the back at waist level; and the grey-blue plumed cap had a decoration of twisted thick gold cord.

The officer with me lamented that all that decoration had disappeared in present-day RAF uniforms. Even the buttons had been reduced from seven to four. 'I have been in the RAF for nearly thirty years, starting off in the ranks, and even the Other Ranks in the Army have their nice gold braid down their trousers,' said the veteran officer. 'We are stuck with our more functional uniform all the time.'

It is rather ironical that the RAF musicians, who used to have plain peaked caps, are now just about the only element of the RAF to display a bit of twisted gold cord – and that as a comparatively new thing. It is certainly successful in winning the attention of the public on great and lesser occasions.

And that is not a matter of merely minor importance to the public's idea of the worth and appeal of the RAF. Even with its image as a high-glamour, high-technology force, the RAF encounters a fickleness in public interest in times of what passes for peace. One of the barometers of this volatility in public understanding and appreciation is the RAF Museum at Hendon, North London, a back-of-beyond which is a substantial distance from the nearest underground station and therefore only to be visited by individuals who are really interested.

'If people are feeding the pigeons in Trafalgar Square and it comes on to rain, then they will go into the National Gallery, even though they may not be interested in art,' the education and public relations officer of the RAF Museum told me. 'There is none of that with us: if people come, it is because they really want to. I think it is probably true that the public *are* fickle. There is a certain proportion of them who are deeply interested, but they naturally enough aren't going to come here month after month. Unless we do something special, like opening a new wing, we get a gradual tapering off of attendances.'

The attendance figures bear this out. In 1973, 550,000 turned up, in 1974 only 350,000 made the journey, and it was not until 1979, the first full year following the opening of the new Battle of Britain Museum, alongside the main museum, that attendances hit the 600,000 mark – an unbeaten record as at 1987. During the years in between, there was a slow slackening off, redeemed only partially in 1983, when the Bomber Command Museum opened on the same site as the other museums, boosting the attendances from 395,000 in 1982 to 495,000 the following year, when the memory of the Falklands may have helped. (It didn't help in another direction crucial to the museum. Although the museum had been nominated by the Ministry of Defence as the charity to which the profits of public events were to be donated that year, almost all events saw about fifty airmen holding out collection boxes for the South Atlantic Fund, which the public understandably found more attractive than other collection boxes.) In 1986, attendances had slumped back to 325,000 again.

Even if numbers are sometimes thin, this is partly redeemed by the intense knowledge of those who *do* go. One man wrote in criticizing the slogan on the body of a Lancaster bomber, ironically quoting Air Marshal Hermann Goering's famous remark that 'No enemy plane will ever fly over Reich territory.' As the complainant was German, at first the museum authorities thought they might have committed a diplomatic error, and braced themselves to argue that the message had been painted on the Lancaster at the same time as it was bombing Germany, and had *not* been added later by the museum. But further down the letter it became apparent that what the German was complaining about was that the quotation was not complete: what Goering had said was, 'No enemy plane will ever fly over Reich territory *or my name is Hermann Müller*.' The German thought the RAF had let the Nazi leader partly off the hook by not recording his rhetorical willingness to be called by the name of a Jew.

Even schoolchildren can be sharp. One boy demanded to know why the missile Blue Steel was painted white. Children and adults share a liking for getting as near as possible to aircraft in the museum; but here adults have the advantage. Staff say that if all the people who ask to climb into rear gunners' cockpits because they were rear gunners themselves in the Second World War had actually *been* rear gunners, the war would have been won in half the time. There is a strict rule against getting into any of the sixty planes: too much damage could result.

The museum has a firm claim to public interest, even if it does not always receive it at a time when the Falklands, let alone the Second World War, may be a fading memory to voters. When I visited it, its very constitution was a muddle of delightful British compromise, each of the three museums on the same site having its separate trustees, who all happened to be the same people, drawn from the RAF, the aeronautical industry and other interested bodies. The reasoning was that it was not suitable for all three museums to be merged until the Bomber Command Museum had discharged its debt (standing at £1,400,000 when I visited the establishment). Since then the museums have merged, and the situation is, I was told, 'much changed'.

Despite the difficulties in Ministry of Defence book-keeping, the museum site has a distinguished history. In 1915 the Royal Naval Air Service put three hangars alongside the runway of Hendon Aerodrome. One of them burned down in the Second World War; but the RAF was still flying from Hendon until the early 1960s, when what had been the airstrip was sold to the Greater London Council for the building of the Grahame Park Estate, the largest council estate built by the GLC. In 1966, the museum was given the two remaining hangars at the instigation of Sir Dermot Boyle, a previous Chief of Air Staff, who became chairman of the trustees. The gap between the two hangars was bridged over to make one large aircraft storage area and an arched façade was fitted to accommodate gallery space and offices.

Outside the museum, in view of the gate, is the bait of a huge and well-preserved RAF Transport Beverley, the predecessor of the Hercules. The oldest plane inside is the Bleriot 11, identical with the one in which Bleriot first flew the English Channel in 1909. The most up to date is an Avro Vulcan of the type that carried out the trip to the Falklands and back, the longest operational flight in history, needing several refuellings there and back without touching down at any fuelling points on land.

Even with such attractions that might be thought to be of relevance to every schoolboy, it is estimated that 50 per cent of those going to the museum are older people on a nostalgia trip, having themselves been involved with the RAF in some way. The museum sends out a lot of literature to schools, and its officials are often asked to give talks, but it often happens that in the teachers' common rooms, before and after the talks, the teachers who make the most effort to talk to the museum's officials are those who argue the pacifist case. The officials have sharpened up their response to this, often arguing that the people of the RAF, rather than the disarmer, are truly pacifist, in the sense that they are doing more to prevent a war.

But in peacetime, it is probably realistic to expect that most people interested in the RAF are those who have some experience of it, either themselves or through relatives. Recently this was borne out by two incidents in succession. In the first, an official was told there was a man at the reception desk with a pilot's medal. The man was a solicitor, clearing up the estate of a client who had died and willed his son's Distinguished Flying Cross to the RAF Museum. The son, a pilot officer during the Second World War, had been with a squadron aboard the aircraft carrier HMS *Courageous* when it was sunk. He died. His father kept the medal until *he* died. Shortly afterwards, the Pilot Officer's sister came to the museum with a clock he had taken from one of the aircraft on Norwegian ice floes which he had been told to set fire to so that they should not fall into German hands. He had taken the clock out of the plane and put it in his luggage for the return journey home; but as his luggage went on a different ship, it survived when he perished.

The second incident involved the De Havilland DH9A, a bomber recovered from Poland in the 1960s in exchange for a Spitfire that the Poles valued more, as many of their countrymen had flown Spitfires during the war. The RAF restoration workshops at Cardington in Bedfordshire spent almost five years rebuilding the DH9A, but were unable to find the controls from one of the dual seats, which had been made detachable so that the second member of the crew would not have the controls waggling about near his knees when he had no need to use them. The museum reconciled itself to putting the DH9A on display without this set of controls. But a few days before this was due to happen, an elderly lady was stopped as she tried to walk through the security gateway of the museum, carrying some strange object under her arm.

'I'm told this is part of an old aeroplane,' she explained. It was a control stick identical with the one missing from the museum's DH9A.

Despite such interest from stalwarts, it is a fact that, after the 50 per cent of visitors who are on a nostalgia trip, there are another 25 per cent who come to study what they consider to be past history rather than to get to know about the RAF in the present (and this despite the fact that the final gallery in the museum, devoted to the present-day RAF, is run by the RAF Director of Recruiting with a rather more generous budget than the more historical exhibits receive). This 25 per cent includes school parties. The remaining 25 per cent of visitors are not, say the staff, so much concerned with the nature or peacetime purposes of the RAF as with its exotic hardware.

'They want to see the hardware like train spotters want to see and take the numbers of trains,' said one member of the staff. 'They collect serial numbers in the same way that people collect engine numbers. The personalities involved don't interest them to a great extent.'

What does interest many visitors in a feminist age is the comparative lack of attention paid to women – though it would be possible to argue that even men cannot compete for attention with the machines. The museum has tried to proof itself against this criticism by having its real-size tableau, showing living conditions in the Second World War, devoted to a WRAF's accommodation hut. One WRAF is shown smartly uniformed in a corrugated-iron Nissen hut; while another, in passion-killing winceyette pyjamas, flops face-down on a tubular portable camp bed. There is a great box of coal next to the free-standing cast-iron room heater, and an iron-legged table and some iron-legged benches. It would be interesting to see how many militant feminists would fare in such an equal environment as that enjoyed by men.

But the issue remains a sensitive point. 'If we want to show a despatch rider, for instance, we try to make it a woman despatch rider rather than a man,' a member of the staff admitted. By contrast, RAF education and the chaplaincy services, for instance, get the brush-off as minority activities: neither gets even a mention in the entire museum.

The museum's director at the time of my visit, Dr John Tanner, an academic and former librarian and tutor at the RAF College, Cranwell, told me his impression was that most people who came to the museum were more interested in aviation than in defence as such – which suggests either a peace-loving nation or a complacent one, depending on the views of those who are judging.

'How many people are interested in defence?' said Dr Tanner. 'Quite frankly, not many. Everyone wants to be defended, and a lot of people are concerned that we don't leave ourselves open to attack ever again. But I don't think defence, as a subject, is one that grasps our public. It is perhaps the history of what the RAF has achieved from World War One onwards that interests people; maybe the interest in the RAF post-1945 is less keen than interest in the pre-1945 era. I would have to say that interest is partly a nostalgic thing, though the gallery devoted to the RAF today is very popular.'

The fact remains that the RAF Museum is one of the best-attended military museums anywhere in Britain or, for that matter, in the world. 'But whether that is because the public is still vitally interested in the RAF as it is today; or whether they are on the whole interested in aircraft and aviation generally; or whether a large body of people still have either a scholarly or romantic interest in the two world wars, I can't say,' admitted Dr Tanner. 'I suggest it is probably a mixture of all these factors.'

I asked whether, in his opinion as director, a large propor-

tion of visitors were motivated to turn up merely because o
a past personal or family connection with the RAF. Dr
Tanner said that was *not* his personal impression, though
the fact that three million people had been in the RAF
during the Second World War did suggest that many, if not
most, present-day families had some family connection with
the RAF.

The fact that the museum has never been a target for
anti-war protesters is probably to be explained by two factors
The first, of course, is that the site is rather inaccessible
The second is almost certainly that the museum does not
brood on the gorier side of war: I did not see one picture
showing people being injured in battle, like one could see
for example, in the Imperial War Museum. Dr Tanner
pointed out that 'We cover the whole history of aviation
starting with mythology and ending up with Concorde
and are not an overtly weapons-brandishing military
museum: we are a little different in tone. Many people have
expressed surprise at the beauty of the museum and the
display techniques. Most of the comments are commendatory rather than condemnatory.'

It seems odd, but true, that the romance of flight may
blind some of the public to the serious purpose of the RAF
as a potential defender in the present day. But the fact that
the glamour can frequently be *seen* by the general public –
unlike the ships of the Royal Navy which are not seen by
most voters from one election to another – undoubtedly
strengthens the public's personal interest in the RAF.

It is ironical that the Americans, who have often gone
their own way in arranging working methods in their Air
Force, sometimes tend to copy social practices which could
derive not merely from the British Royal Air Force but from
the British Army before that. It comes as a surprise to
United States Air Force officers on exchange to Britain when
they come across the RAF tradition of dining-in, which
means attending dinner in mess in full dress uniform once
every so often – in these days, usually no more than once a
month. But it is less of a surprise than it would have been a

decade ago, for the Americans are themselves concocting mess rituals as a way of boosting cohesion and morale.

One American Air Force officer on an exchange tour in Britain said: 'We have made our own adaptations and changes to it. We have Americanized it. In Britain it is always the Station Commander or Officer Commanding who takes the chair at these events, whereas we Americans often rotate the presidency, so that different commanders can have a turn.'

In Britain there was a functionary at these functions called 'Mr Vice', who in effect was an officer of lower rank through whom toasts were formally channelled – and that was just about the end of Mr Vice's activity. Usually Mr Vice was a Lieutenant or Captain in the American version; and he played a much more active role.

How so? I asked the United States Air Force Captain.

'Well, another custom we have established is that, if a person making a toast makes a mistake, stumbles, someone can ask that he goes to the grog bowl. This is a very evil liquid in a container, which can be almost anything. The Sergeant at Arms will be requested to march the offender to the grog bowl and see that he drinks a cup from the bowl. Usually it is not very pleasant. Usually there would be alcoholic and non-alcoholic components. It could be anything Mr Vice and the committee decide to put into it. That is part of the active role of Mr Vice.'

Apparently on one occasion, Mr Vice and the committee decided to get a lavatory bowl, sanitize it and fill it with a nauseating mixture – with vegetables floating on the top of it which could suggest some other sort of substance. The unfortunate officer found himself drinking a mixture of hard liquor, liqueurs and wine. It was his *going-away* dining-in – for which, I would have thought, he might have been grateful.

I did not come across the exact equivalent of this in the RAF, though young pilots and navigators sometimes celebrate the end of training flights to Germany and elsewhere by preparing drinking mixtures which tasted to me like cocktails made with eau de Cologne, gasoline and paint-stripper.

As a civilian, I was greatly relieved that I didn't have to face the American version, and drank the Royal Air Force's mixture as a gesture of comradely thanks.

In Conclusion

As the youngest of the fighting forces, the RAF has not had time to become encased in historic but perhaps restricting traditions.

The weapons it handles have always been so seriously destructive that the quaint initiation ceremonies which in the Army are supposed to demonstrate the 'toughness' of the new recruit have generally appeared redundant.

About a third of the RAF's officers still come from public schools, but a third have come up through the ranks. And it is far more difficult to tell from the cast of face and personality which is which – rather more difficult than it might be in the case of even the modern British Army or Royal Navy. One Warrant Officer I spoke to in the intricate labyrinths of the secret communications centre at Cyprus, a bald and avuncular man, could easily have been a Wing Commander. Very many Wing Commanders I met could have been Warrant Officers – indeed, many *had* been.

When visiting the Royal Navy for *Ruling the Waves*, it was obvious that some experienced non-commissioned officers somewhat resented the arbitrary way they felt they were constantly being overruled by young and inexperienced officers (and that their wives resented it on their behalf even more). In the RAF I did not come across the same divisive syndrome at the professional level.

But there was a feeling there still remained a *social* divide between commissioned and non-commissioned men and their wives; and that friends tended to keep their distance once the commissioned/non-commissioned divide got between them.

This social cost may be paid increasingly grudgingly by a

population that has known only European peace for the past forty years; but it is widely recognized that the social divide may be inevitable in any disciplined and armed service. The simple example always given (on both sides of the social divide) is that an officer could not easily, on the morning after a convivial binge with a non-officer, send the man out on a mission that might well endanger his life; and that officers must remain objective and therefore slightly distant about the men under them *and* their families. This explanation seems to be generally accepted by both sides.

The limitations imposed on women in their life in the RAF are less universally defended. Women are now trained in the use of automatic weapons as well as hand guns, but only for guard duties and personal protection. They cannot reach the highest ranks in the RAF: those ranks demand combat training and experience, and women are still not allowed to become pilots, navigators or anything else actually in the air except air loadmasters on freight flights.

A large proportion of members of the Women's Royal Air Force I spoke to said that if they were required to drop bombs on people they would be tempted to leave the service. A few did say they would be prepared to be *fighter* pilots, because they saw this as an essentially defensive, life-*saving* role: the defence of Britain and British lives.

The position of women in the RAF is thought unlikely to expand in the next few years. That being so, it is easy for some women to say they would dearly love to be pilots or navigators in charge of aircraft that may have to fire or be fired upon; they know they are not likely to have their keenness tested. And those who openly declare they would *not* be prepared to enter combat, as pilots or navigators, know that there will be no social or professional pressure on them to do so.

If there is a case for a more classless and sexually non-discriminatory armed service, then the RAF, as the most recently formed and least hidebound of the three services, would seem to be the most logically poised to become it.

But from the very first, officer recruits are treated differently from others; and men are treated differently from women.

Cranwell, where officers are trained, has the atmosphere of a great public school. Swinderby, where the airmen are trained, only a few miles up the road in Lincolnshire, is a haphazard collection of utilitarian buildings that together suggest a bare transit camp where young men are routinely shouted at, much as they might be in the Army. There is arguably some room for reform here, so that the message to new recruits is: 'We have high standards of environment and behaviour here, and we expect you to rise to them or get out.'

Conditions for girls in the junior ranks are, in practice, more comfortable and rational than those enjoyed by the men. Rooms are often better appointed. Officers will explain to you that girls are rarely shouted at because they do not respond well to being shouted at. Do the more intelligent boys, one wonders? Some are taken aside by instructors who diplomatically whisper in their ears that this is just initial training: 'It won't be like this when you actually get into the RAF proper.' Could it be that there is a case for making comparatively civilized conditions for both sexes in the RAF the keynote right from the start? In days of high unemployment in civilian life, the RAF could afford to jettison those men who did not respond to more sophisticated treatment.

At present, as with the other two services, conditions are deliberately arranged so that the best environment goes to those with the highest rank. This, it is thought, is an incentive to ambition. But whether the differences need to be so extreme is arguable: in civilian life a de luxe model of a car instead of a basic one, or an extra square foot of office carpet, is an incentive that is eagerly intrigued for – very satisfactorily from the employer's point of view, since it costs comparatively little.

On the other hand, the RAF top brass are eager for the RAF to be seen as a fully fledged *fighting* force no less combat-worthy than the Army or Royal Navy. There is no more chance today of having a Supply Branch officer as Marshal of the Royal Air Force than there was when the RAF was formed on 1 April 1918. And there seems no chance, either, of the junior ranks being allowed either alcohol or the opposite sex in their own rooms. This rule, which may sound feudal to a civilian, was unhesitatingly defended by all officers with whom I raised the subject. The line invariably was: 'We don't *need* to try relaxing the rules to see what would happen: we *know* what would happen from experience. It is a recipe for disaster. And you must remember that with some of the younger boys and girls, we are responsible to their parents. Basic human nature doesn't change with the times.'

Quite so. In many ways, the RAF is changing as society is changing, but it still treads a delicate path between the cerebral requirements of high technology and the rigid discipline upon which lives may depend in a fighting force. In that sense, it probably has a greater degree of internal tension than the other two forces I have dealt with in *The People of the Forces Trilogy*, and which are equally crucial to our national safety.

Men who are thought fit to fly fighters worth £20 million, who have to exercise a degree of personal decision-making worthy of a highly paid executive in civilian industry, are not necessarily those most at home with a life-style dictated by military rules and customs. There have been signs that pilots, and others with skills that can be imported into civilian life, have not been altogether happy with their service lives. Perhaps it is to be explained by the fact that all services' personnel are liable to be moved around with more arbitrariness than civilians, a fact at odds with the understandable desire of RAF men and women to get a toehold in the rising property market by taking out a building society mortgage on a permanent home. Perhaps it is just that, in an era when there seems no immediate threat to the United Kingdom, the various disciplines of service life tend to grate more than they would do if a Warsaw Pact invasion of Europe seemed more imminent, as one day it might be.

Whatever the cause (and people are not always clear themselves about their precise motives), the House of Commons was told, at the beginning of 1987, about what was termed a 'black hole' opening up in the RAF: 175 pilots had left it in the previous year, 114 of them without completing their periods of engagement. A further 213 had expressed a wish to leave, 184 of them not having completed their engagements. Many had left to get large salaries in Saudi Arabia; others had gone to British civilian airlines.

Civilian employment has seemed far from secure and predictable in the 1980s. It is understandable that pilots (who would have to retire at fifty in any case, and might, in RAF terms, be thought to be past their best at thirty-five) might want to take safe pensionable jobs in civilian airlines immediately the opportunities occurred, rather than wait till the end of their service contracts in the mere *hope* that jobs might be available then. It is equally understandable that the attractive pay available in the Middle East to many skilled people, from footballers to nurses, would attract some RAF pilots.

But the drain of pilots, each trained at £3 million a time or more, cannot solely be put down to transitory circumstances. It may reflect the view of some RAF men that, though the RAF has moved with the times, in some ways it may not have moved far enough and fast enough.

An awareness of this in the upper ranks of the Royal Air Force – and I suspect the awareness of the need to meet changes in society is always there – makes the safety of our skies more secure as we move towards the twenty-first century and a greatly changing international scene.